TRACKING
THE
AUTOMATIC
ANT

Springer
New York
Berlin
Heidelberg
Barcelona
Budapest
Hong Kong
London
Milan
Paris
Santa Clara
Singapore
Tokyo

DAVID GALE

TRACKING THE AUTOMATIC ANT

AND OTHER MATHEMATICAL EXPLORATIONS

A Collection of Mathematical Entertainments Columns from

The Mathematical
𝕴𝖓𝖙𝖊𝖑𝖑𝖎𝖌𝖊𝖓𝖈𝖊𝖗

Springer

David Gale
Department of Mathematics
University of California
921 Evans Hall
Berkeley, CA 94720
USA

Appendix 1—A Curious Nim-Type Game, reprinted from the *American Mathematical Monthly,* Vol. 81, No. 8, October 1974, pp. 876–879.

Appendix 2—The Jeep Once More or Jeeper by the Dozen, reprinted from N.J. Fine's *The Jeep Problem, American Mathematical Monthly,* Vol. 54, 1947, pp. 24–31.

Library of Congress Cataloging-in-Publication Data
Gale, David.
    Tracking the automatic ant and other mathematical explorations / David
Gale.
        p. cm.
    ISBN 0-387-98272-8 (alk. paper)
    1. Mathematical recreations.   I. Title.
    QA97.G127   1998
    793.7'4–dc21   97-33272

Printed on acid-free paper.

Production managed by Timothy Taylor; manufacturing supervised by Jeffrey Taub.
Composition by Impressions Book and Journal Services, Inc., Madison, WI.
Printed and bound by R.R. Donnelley and Sons, Harrisonburg, VA.
Printed in the United States of America.

9 8 7 6 5 4 3 2 1

ISBN 0-387-98272-8 Springer-Verlag New York Berlin Heidelberg   SPIN 10557669

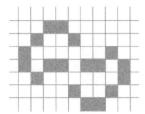

This book consists mainly of material from the column *Mathematical Entertainments* in *The Mathematical Intelligencer,* a feature I edited from 1991 through 1996. The column had been devoted almost entirely to problems, but when I started as editor, I was encouraged by the editor of the magazine, Sheldon Axler, to do with it whatever I pleased. As a result the problems became fewer and fewer and eventually disappeared altogether. Instead I was able to write about anything that caught my fancy. As a result the topics taken up vary all over the lot, but almost always are concerned with very elementary things. Thus three of the chapters are devoted almost entirely to triangles, two others to tiling by rectangles, three more to mysterious properties of sequences given by simple recursions, three to games and paradoxes, and three to a particular automaton.

As can be seen from this listing, there seems to be no unifying thread to the collection, but I did follow some general principles in deciding which material to use.

(1) Roughly speaking, one should not have any specialized knowledge in order to understand the

chosen topics (though occasionally the subsequent analysis becomes a bit more advanced).

(2) A high priority was placed on mathematical surprises and unexpected twists, some of which have explanations, while others remain mysterious.

Both of these principles are illustrated by examples from everyday life that turn out to have unexpected mathematical content. Among these are familiar children's games and two sections on tying shoelaces. In the same spirit, in a section on geometric constructions, it turns out that one can throw away the unwieldy compass and use instead a straightedge on which one can make marks. This apparatus allows one to make not only the Euclidean constructions, but also any construction involving equations of degree at most four, thus making it possible to trisect angles and duplicate cubes (see Chapter 10). In the first three appendices, I present the original sources for some of the material.

Computers play a role in many of the sections, but in various guises. Sometimes they are used simply for performing computations, such as finding the power of 2 whose first four digits are 1492 (Chapter 7). Elsewhere they are used for experimentation and, perhaps most interestingly, for proving theorems. For example, one obtains dozens of new theorems of plane geometry by discovering numerically that certain geometrically defined points are collinear (Chapter 6). When these discovered collinearities are then formulated algebraically, the computer again steps in, in its symbolic rather than numerical mode, and provides the proofs.

From the point of view of this book, however, the most intriguing consequence of the use of the computer is how it affects the way we think about mathematics itself. Traditionally, mathematics has been *the* example of a deductive science. Thus, no matter how much "experimental evidence" one has for, say, the truth of Fermat's famous theorem, it isn't considered mathematics until it has been proved, that is, shown to be deducible from previously known results. But things started to change in 1931, when Kurt Gödel showed that there is no nontrivial mathematical system that can prove or disprove all the statements that can be made in that system. Today, by enabling us to amass enormous amounts of empirical data, computers have thrown a spotlight on the crisis. For example, at latest count we now know several billion digits of π, and one-tenth of them are 7s to within a very small epsilon. Yet no one has any idea how to show that this behavior goes on "forever." Indeed, for all we know, there may be no 7s at all from some point on. Optimists would say that there has to be a proof out there somewhere, and when we become sufficiently clever, we will find it. But what if there is no proof? In Chapter 3 and 5 we look at other types of questions about sequences where certain results seem overwhelmingly likely yet the prospects for finding proofs seem overwhelmingly remote. Thus, mathematicians find themselves in a curious situation. At one extreme, by amassing data computers can first supply us with conjectures and then go on, using programs we design, to settle them. This is the case with the geometrical results mentioned above. At the other end, the computer may be giving us glimpses of true mathematical facts for which there is no proof. Of course, between the extremes there is enough room to keep mathematicians busy for the foreseeable and even the unforeseeable future.

So the advent of computers has made mathematics more like the physical sciences, in that it now has a strong experimental as well as theoretical component. To find out what is going on in the physical universe, physicists use particle accelerators, and astronomers use high-powered telescopes. For mathematicians, the computer now plays an analogous role. The important difference is that the computer does not give us information about the physical universe but rather about the "abstract" universe. The things we study, finite groups, analytic functions, and the like are no less "real" than atoms or galaxies. The difference is that the objects are abstractions, things that wouldn't be there if we hadn't invented them. Of course many of these abstractions were originally created in an attempt to understand the physical universe, the prototypical example being the books of Euclid. But there is also so-called "pure" mathematics, in which the objects of study are the abstractions themselves, and herein lies a great paradox. As an illustration let us look again at the digits of π and see what is involved in describing the problem. First of all we need the notion of a circle. Nature offers suggestive examples: the sun, the moon, the head of a mushroom. But nature does not supply the center. We had to invent that, likewise the circumference and the diameter, the concept of length as a "number" (which we also invented), and finally decimal expansions. All of these things are creations of the human mind, and yet they seem to possess a life of their own, as they rise up and present us with puzzles, some of which we may never be able to solve. The last appendix to the book explores further the philosophical implications of this state of affairs.

Finally, a word of clarification. As noted, the columns were originally published as "entertainments," and I have referred above to mathematical phenomena as providing us with puzzles. The everyday connotation of these words suggests that one is here concerned only with lightweight matters as opposed to "serious mathematics." This distinction is more apparent than real. In fact, one might say that the whole scientific enterprise is about solving the puzzles nature keeps throwing at us. There is, however, one sense in which I have tried to keep the presentation on the light side. The process of trying to understand mathematical results can be very hard work. By contrast, in this volume I tried to choose topics from which readers could reap the rewards of discovery with at most a moderate expenditure of energy, in other words, to maximize the ratio of pleasure to effort. Although the book does not attempt to read like a novel, my hope is that in dipping into its pages, the reader will find things that are entertaining, yes, but also illuminating.

University of California   David Gale
Berkeley

Acknowledgments

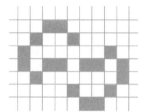

I would like to acknowlege contributions from people who played important roles in helping to put together the material for this book.

I am grateful for the encouragement and support of Sheldon Axler and Chandler Davis, the two editors of *The Mathematical Intelligencer* who gave me free hand to write about whatever caught my fancy during my time as editor of the Mathematical Entertainments column. Some of the chapters were written in part or entirely by "guest" columnists, including Donald Newman, Jim Propp, Scott Sutherland, Serge Troubetzkoy, Sherman Stein, John Halton, Solomon Golomb, Michal Misiurewicz, and Armando Machado.

In addition, material for many of the chapters was obtained from discussions, email messages and telephone conversations with many people, among them Michael Somos, Imre Barany, Clark Kimberling, Elwyn Berlekamp, Joe Keller, Clifford Gardner and especially the late Raphael Robinson who made key contributions to almost all of the early chapters.

All of these people have made the writing of the columns and the resulting book a sort of cooperative venture. I hope the readers will now share some of the pleasure that we had in putting this book together.

Contents

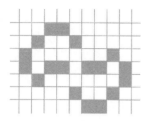

Preface   *v*

Acknowledgments   *viii*

CHAPTER 1
Simple Sequences with Puzzling Properties   *1*

CHAPTER 2
Probability Paradoxes   *7*

CHAPTER 3
Historic Conjectures: More Sequence Mysteries   *13*

CHAPTER 4
Privacy-Preserving Protocols   *19*

CHAPTER 5
Surprising Shuffles   *25*

CHAPTER 6
Hundreds of New Theorems in a
Two-Thousand-Year-Old Subject:
Where Will It End?   *31*

CHAPTER 7
Pop Math and Protocols   *37*

CHAPTER 8
Six Variations on the Variational Method   *45*

CHAPTER 9
Tiling a Torus: Cutting a Cake   *55*

CHAPTER 10
The Automatic Ant: Compassless Constructions   *63*

CHAPTER 11
Games: Real, Complex, Imaginary   *73*

CHAPTER 12
Coin Weighing: Square Squaring   *83*

CHAPTER 13
The Return of the Ant and the Jeep   *91*

CHAPTER 14
Go   *101*

CHAPTER 15
More Paradoxes. Knowledge Games   *113*

CHAPTER 16
Triangles and Computers   *119*

CHAPTER 17
Packing Tripods   *131*

CHAPTER 18
Further Travels with My Ant   *137*

CHAPTER 19
The Shoelace Problem   *151*

CHAPTER 20
Triangles and Proofs   *163*

CHAPTER 21
Polyominoes   *171*

CHAPTER 22
A Pattern Problem, A Probability Paradox, and A Pretty Proof   *189*

CHAPTER 23
The Sun, the Moon, and Mathematics   *199*

CHAPTER 24
In Praise of Numberlessness   *205*

APPENDIX 1
A Curious Nim-Type Game   *213*

APPENDIX 2
The Jeep Once More or Jeeper by the Dozen   *217*

APPENDIX 3
Nineteen Problems in Elementary Geometry (by Armando Machado)   *227*

APPENDIX 4
The Truth and Nothing But the Truth   *233*

chapter 1

Simple Sequences
with Puzzling
Properties

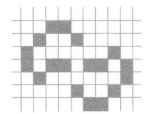

COMPUTER-GENERATED MYSTERIES

Many mathematicians feel that the main impact of computers on mathematics has been not in solving problems, as one might expect, but rather in posing them. The prime illustration is probably the recent activity in discrete dynamical systems stimulated by the celebrated computer experiments of Mitchell Feigenbaum. Perhaps "explorations" is a better description of this work, for the appropriate analogy is not with physics or biology but with astronomy. The computer is the mathematician's telescope, which when used intelligently makes it possible to find out what is "out there" in the mathematical universe. (This whole development should be a source of satisfaction to the Platonists, who have been saying all along that, like stars and galaxies, mathematical phenomena are discovered, not invented.)

The recent work described here gives another striking example of a set of phenomena that would probably never have been observed without the use of computers.

THE STRANGE AND SURPRISING SAGA
OF THE SOMOS SEQUENCES

In investigating properties of elliptic theta functions, Michael Somos discovered an infinite sequence whose first 15 terms are

$$1, 1, 1, 1, 1, 1, 3, 5, 9, 23, 75, 421, 1103, 5047, 41783.$$

The sequence is defined by $a_i = 1$ for $0 \leq i \leq 5$ and

(1) $$a_n = (a_{n-1}a_{n-5} + a_{n-2}a_{n-4} + a_{n-3}^2) / a_{n-6} \quad \text{for } n > 5.$$

The surprising fact is that the recursion generates integers as far as the eye = computer = telescope can see. In fact, for this example a telescope is not required. A good pair of binoculars will do. With a pocket scientific calculator one easily, for example, verifies that the next numerator above is divisible by 23, so that a_{15} is again an integer. What's going on?*

Upon observing this phenomenon, it occurred to a number of people to consider the simpler fourth-order recursion

(2) $$a_n a_{n-4} = a_{n-1}a_{n-3} + a_{n-2}^2, a_0 = a_1 = a_2 = a_3 = 1.$$

Once again all entries turn out to be integers, but in this case the situation is manageable, and several people have come up with proofs, the first one given by Janice Malouf. We present here a variant due to George Bergman. First note that because of (2) every four consecutive terms of the sequence are pairwise relatively prime. For suppose this is true up to a_n. Then a_n would have a prime factor p in common with a_{n-1} or a_{n-3} if and only if p also divided a_{n-2}, contrary to the induction hypothesis.

We now show inductively that if $a_{n-4}, \ldots, a_n, \ldots, a_{n+3}$ are integers (clearly true for $n = 4$), then so is a_{n+4}, and hence all a_i. Writing $a_{n-3} = a, a_{n-2} = b, a_{n-1} = c$, we have $a_n a_{n-4} = ac + b^2$, so a_n divides $ac + b^2$. By the preceding paragraph we may apply (2) to the sequence modulo a_n, giving

(3) $$a, b, c, 0, \frac{c^2}{a}, \frac{c^3}{ab}, \frac{c^3}{a^2}, a_n a_{n+4} \equiv \frac{c^5}{a^3 b^2} (ac + b^2) \equiv 0.$$

So a_n divides $a_n a_{n+4}$.

Although the proof is very simple, it depends on the fortuitous fact that the factor $ac + b^2$ turns up on the fourth iteration. We will return to this point.

The same method works for the five-term recursion

(4) $$a_n a_{n-5} = a_{n-1}a_{n-4} + a_{n-2}a_{n-3}.$$

Actually, in all of these recursions one may put arbitrary integers as coefficients of the terms $a_{n-i}a_{n-j}$ and still get integers, and this can be proved for the recursions (2) and (4).**

---

*Somos actually discovered his sequences 14 years ago but did not succeed in capturing the attention of the mathematical community until the summer of 1989.

**If the integer coefficients are allowed to be negative, then it may happen that some $a_n = 0$, in which case we shall adopt the convention that the sequence terminates at that point.

The next bit of progress came when Dean Hickerson proved that the original Somos sequence gives integers. In fact he showed something more general. Instead of starting with six one's he considered the sequence starting with indeterminates $a_0, a_1, \ldots, a_5$. The recursion then generates rational functions $a_n = p_n/q_n$ of these a_i, and the theorem is that the denominators of these functions are always monomials with coefficient 1. This is clear for $a_0, \ldots, a_{11}$, but note that to compute a_{12}, one must divide by $a_6 = (a_5a_1 + a_4a_2 + a_3^2)/a_0 = p_6/a_0$. Hand computation of a_7, a_8, etc. quickly becomes unmanageable. This is what symbolic manipulation programs are designed for, and using *Macsyma*, Hickerson found that as in (3) above, p_6 occurs as a factor of the numerator of a_{12} (which when reduced to lowest terms contains 194 terms!). Further *Macsyma* calculations are used to prove that p_6 is prime to $p_7, \ldots, p_{12}$, and an inductive argument is used to complete the proof. (Richard Stanley has also solved this problem using similar methods.)

But what have we learned? As Hickerson puts it, "The thing I dislike about my proof is that it doesn't explain why the result is true. It depends primarily on the fact that when you compute a_{12}, there's an unexpected cancellation. But why does this happen?" Indeed, the proof, rather than illuminating the phenomenon, makes it if anything more mysterious. I report this with some embarrassment, since I have earlier asserted that a proof in mathematics is in some sense equivalent to an explanation. We now see that this clearly need not be the case. Perhaps, if and when we find the "right" proof, the situation will become clarified, but must there necessarily be a right proof? One is reminded of the proof of the four-color theorem. One of the most interesting features of the Somos problem is that it leads to this sort of speculation.

Getting back to the question at hand, having found proofs for recursions of order 4, 5, and 6 and empirical evidence for 7, we find that those of order 8 and above do not give integers. You can easily confirm with your pocket calculator, for example, that for the recursion of order 8, a_{17} is a fraction. Curiouser and curiouser.

The next discovery is due to Raphael Robinson, who found that the integer property of recursions (1), (2), and (4) was (apparently) shared by an infinite family of recursions. For any $k \geq 6$, start with k ones and then use the recursion

(5) $$a_na_{n-k} = a_{n-1}a_{n-k+1} + a_{n-2}a_{n-k+2} \quad \text{or}$$

(5') $$a_na_{n-k} = a_{n-1}a_{n-k+1} + a_{n-2}a_{n-k+2} + a_{n-3}a_{n-k+3}.$$

The fact that one is now dealing with an infinite collection of sequences would seem to put the problem out of range of *Macsyma*-type proofs.

At this point my pocket calculator convinced me that for any $0 < l < m < k$, the recursion

(6) $$a_na_{n-k} = xa_{n-l}a_{n-k+l} + ya_{n-m}a_{n-k+m}$$

gives integers, generalizing (5). Further investigations, again by Robinson, lead to the following:

Conjecture: For any $p, q, r < k$, the recursion

(7) $$a_na_{n-k} = xa_{n-p}a_{n-k+p} + ya_{n-q}a_{n-k+q} + za_{n-r}a_{n-k+r}$$

generates integers if and only if p, q, r are chosen such that $p + q + r = k$.

(Robinson's evidence is only for the case $x = y = z = 1$. The arbitrary x, y, z are my responsibility.) This would subsume (5′) and (6), namely, (6) corresponds to choosing $p = l$, $q = k - m$, $r = m - l$, and $z = 0$, and (5′) corresponds to choosing $p = 1$, $q = 2$, $r = k - 3$, $x = y = z = 1$.

The story is not over. Dana Scott set up a program for the simplest case $k = 4$ but forgot to square the term a_{n-2}. Yet the recursion still gave integers! In fact, it turns out that recursion (2) can be generalized to

$$(8) \qquad a_n a_{n-4} = a_{n-1}^p a_{n-3}^q + a_{n-2}^r \quad \text{for any } p, q, r > 0,$$

and the Bergman proof goes through as it does for recursion (4) with arbitrary exponents. On the other hand, one cannot choose arbitrary exponents *and* coefficients. In fact, the recursion $a_n a_{n-4} = 2a_{n-1}a_{n-3} + a_{n-2}$ does not give integers (although if the right-hand side is $a_{n-1}a_{n-3} + y a_{n-2}$, it can be proved that the recursion gives integers for all y).

Recursion (8) is interesting because in all the other examples the right-hand side was homogeneous. Was this a red herring? Perhaps, but when we go to three-term sequences, we can no longer throw in arbitrary exponents. In fact, if one forgets to square the term a_{n-3} in the original sequence (1), one gets fractions.

Perhaps the simplest recursion of all has been discovered by Scott. For any k,

$$(9) \qquad a_n a_{n-k} = a_{n-1}^2 + \cdots + a_{n-k+1}^2$$

which seems to work for all k. Other "good" recursions seem to be

$$(10) \qquad a_n a_{n-k} = a_n a_{n-2} + \cdots + a_{n-k+2}a_{n-k+1}$$

and, for k odd,

$$(11) \qquad a_n a_{n-k} = a_{n-1}a_{n-2} + a_{n-3}a_{n-4} + \cdots + a_{n-k+2}a_{n-k+1}.$$

These recursions break new ground, since the right-hand side may have any number of terms, whereas in previous examples three terms seemed to be the maximum. For $k = 4$ the Bergman proof works for (9) and (11) but not for (10), which (for the moment) remains unsolved.

I don't want to drag this out indefinitely, for it seems that new examples of Somos sequences are coming in faster than I can write them down. There is a whole area in which one uses recursions like (1) but starts with sequences other than all ones, e.g., ones and twos or ones and minus ones. Experiments indicate that sometimes one gets integers and other times not, but there seems to be no discernable pattern. On the positive side, using the ideas of Hickerson, Gale and Robinson have proved integrality for the sequences (5) (but not (5′)). I strongly suspect that by the time this appears in print much more will be known about Somos sequences. Perhaps the problem will have even been solved, but as of this writing the situation remains intriguingly mysterious.

Addendum. Within weeks after this column was written. Ben Lotto, using the ideas of Hickerson but no computer calculation, showed that (9) always gives integers. The method doesn't seem to work, however, for (10) and (11), although Robinson, using an entirely different but elementary argument, showed that (10) gives integers for $k = 4$, settling the question raised two paragraphs above. Also, conjecture (7) was proved for $k = 7$, $p = q = r = 1$ (using *Mathematica* rather than *Macsyma* this time). Finally, Robinson discovered a whole set of periodicity phenomena that occur when the values of the terms in the sequences are reduced modulo n. Periodicity has been proved for (2), (4), and for (10) with $k = 4$, and for (9) with $k = 3$, but it remains unexplained otherwise.

A THEOREM-JOKE CONTRIBUTED BY HENDRIK LENSTRA

"*Perfect squares don't exist.* Suppose that n is a perfect square. Look at the odd divisors of n. They all divide the largest of them, which is itself a square, say d^2. This shows that the odd divisors of n come in pairs a, b, where $a \cdot b = d^2$. Only d is paired to itself. Therefore the number of odd divisors of n is odd. This implies that the sum of all divisors of n is also odd. In particular, it is not $2n$. Hence n is not perfect, a contradiction: perfect squares don't exist."

Get it?

Remark: It seems the joke works only in English. In other languages a square is just a square (the theorem, however, is international).

IS THERE A MATHEMATICS GENE?

The four-year-old niece of a mathematical logician was playing a game in which she was the conductor on a train and her mother was a passenger. "Wait a minute," said Nancy, "we have to get some paper to make tickets". "Oh", said her mother, who had probably had a long day, "Do we really need them? After all, it's only a pretend game with pretend tickets." "Oh no, mommy, you're wrong," replied Nancy, "They're pretend tickets, but it's a real game."

chapter 2

Probability Paradoxes

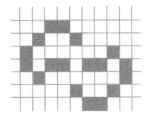

PARADOXES AND A PAIR OF BOXES

What is a paradox? Perhaps the best-known examples in mathematics are Russell's paradox and the Banach–Tarski paradox, but these two results are very different. The Banach–Tarski theorem is considered paradoxical because it shows that sets can behave in a way very different from our intuitive notions about them. Russell's paradox, on the other hand, shows that starting from what seem to be plausible axioms, one can arrive at a contradiction. The proper term for this is not paradox but *antinomy*, which, according to Webster, is "a contradiction between two apparently equally valid principles or between inferences correctly drawn from such principles," whereas a paradox is "a statement that is seemingly contradictory or opposed to common sense and yet is perhaps true."

The first two examples below are antinomies and the last two are paradoxes.

1. *The other box.* You are presented with two boxes, each containing an amount of money that has been placed there by the following rule. A fair coin was tossed until it fell tails. If n heads were tossed, then one

of the boxes contains 3^n dollars and the other 3^{n+1}. You are allowed to open one of the boxes and count its contents. You may then either pocket this money or switch and take the money in the as yet unopened box. What should you do? Clearly, if the box you open contains one dollar you should take the three dollars in the other box. Now suppose the box you open contains 3^n dollars. Then one easily sees that the other box will contain 3^{n-1} or 3^{n+1} dollars with respective probabilities $2/3$ and $1/3$. So your expectation from switching is

$$\frac{2}{3}3^{n-1} + \frac{1}{3}3^{n+1} = \frac{11}{9}3^n > 3^n,$$

so you maximize your expected winnings by switching. Assuming you are an expectation maximizer, this is what you will do. (The debate as to whether expectation maximizing is "reasonable" is of course beside the point, because we are concerned here only with the mathematics and not with its implications for behavior.) But now, because you know in advance that you will always switch, there is no point in wasting time opening and counting. You should simply choose "the other box" to begin with, and the same argument then shows that whichever box you choose, you would have been better off expectationwise to have chosen the other.

This game has a definite Petersburgian flavor in that the expected winnings in the game are infinite. The novelty of this variant is that it seems to lead to a contradiction, and thus we are dealing with an antinomy rather than a mere paradox.

2. *Beat the house.* A somewhat similar example was told to me by Lester Dubins, who is uncertain of its origin. In a certain casino one can play the following game. The house posts a positive integer n. In this game it is you the customer who are invited to toss the fair coin until it falls tails. If you tossed $n - 1$ times, then you pay the house 8^{n-1} dollars, but if you tossed $n + 1$ times, you win 8^n dollars from the house. In all other cases the payoff is zero. Since the probability of tossing exactly n times is $1/2^n$, your expected winnings are $8^n/2^{n+1} - 8^{n-1}/2^{n-1} = 4^{n-1}$, for $n > 1$, and 2 for $n = 1$. So your expected gain, which is the house's expected loss, is positive. But now it turns out that the house arrived at the number n by tossing that same fair coin and counting the number of tosses up to and including the first tails. Thus, you and the house are behaving in a completely symmetric manner. Each of you tosses the coin, and if the number of tosses happens to be the consecutive integers n and $n + 1$, then the n-tosser pays the $(n + 1)$-tosser 8^n dollars. But we have just seen that the game is to your advantage as measured by expectation no matter what number the house announces. How can there be this asymmetry in a completely symmetric game?

3. *The other box again.* (A mild modification of an example due to David Blackwell.) This time the boxes do not contain money, but each box contains an integer (perhaps printed on a card), and the only thing you know about them is that they are distinct. You draw one of them at random and are then supposed to guess whether the other is higher or lower. Is there anything you can do to give yourself a better than even chance of guessing correctly? Surprisingly, the answer is yes, provided that you have some mechanism for randomizing such as a true coin to toss. Suppose you have a spinner like those used in children's board games. You should proceed to spin and then record the angle θ between the initial and final position of the pointer. Now draw your number and guess higher or

lower according to whether cot $\theta/2$ is greater or less than the number you drew. Let us assume for convenience that θ has the uniform distribution on $[0, 2\pi)$. Claim: If the two numbers are $p > q$, then the probability that you will guess correctly is $1/2 + (\cot^{-1}q - \cot^{-1}p)/2\pi$. Namely, $q < \cot \theta/2 < p$ if and only if $2 \cot^{-1}p < \theta < 2 \cot^{-1}q$. From uniformity this has probability $\gamma = (\cot^{-1}q - \cot^{-1}p)/\pi$. In this case, because of the guessing rule, you will be right no matter which number you draw. In the other cases, where θ is either greater or less than both p and q, your probability of guessing right is one-half, since you are equally likely to draw p or q. Thus your probability of a correct guess is $1/2 (1 - \gamma) + \gamma = 1/2 + \gamma/2$, as claimed.

Mathematically, this argument is airtight, but it raises some interesting philosophical questions about the applicability of probability theory in decision-making. For example, suppose you don't have a spinner but are wearing a watch, the nondigital kind, and choose θ to be the angle between the minute and hour hands. Is this in any sense a kind of randomization, and if not why not? Suppose instead of numbers the boxes contain stones of different weights, and you have a balance so you can compare weights but no scale for making measurements. You draw a stone and must guess whether the other is heavier or lighter. Is there anything you can do? This might lead one to think that randomization is possible only for *quantitative* comparisons, meaning that one must associate numbers with the objects drawn. But this need not be true either. Suppose, for example, the boxes contain slices of pie. Then a spinner is just what you need. Guess bigger or smaller depending on whether the spinner angle is greater or less than the angle of the slice you draw. This can be determined by direct visual comparison. There is no need for a protractor, and numbers are not involved in any way.

4. *The other's number.* This time the integers in the boxes are positive and consecutive. Each player draws one and is supposed to find out the opponent's number by the following procedure. The players are equipped with a blank card and a pencil. If at any time a player knows her opponent's number, she writes it on her blank card and wins the game. If neither player knows the other's number, they exchange blank cards and start over. The assertion is that with perceptive players, this game will terminate. More precisely, we state it as a theorem:

Theorem. *If the two numbers are n and $n + 1$, then the player holding n will win after $n - 1$ exchanges.*

The proof is by induction. If $n = 1$, then the player holding 1 will know that the opponent's number is 2, and the game ends with no exchanges. Now assume that the conclusion is true up to n, and suppose the lower number is $n + 1$. Then the player holding this number knows that if her opponent holds n, he will end the game after the $(n - 1)$st exchange (induction hypothesis). So when he doesn't do this, she knows after the exchange that he must hold $n + 2$, and she wins.

The paradoxical point is this: suppose the numbers held are, say, 72 and 73. Then neither player knows the other's number, and both are aware of their opponent's ignorance. So they know for certain that the first stage of the game will be an exchange of cards. When this indeed takes place they have apparently gained no new knowledge. Yet since the game is now one step closer to termination, something must have changed. What was it?

WE ALL MAKE MISTAKES

Some of us more than others, so it may be comforting to realize that occasionally even the great mathematicians have published erroneous results. Readers are invited to contribute examples of this phenomenon, where a mistake should mean not just a gap in a proof or a case of meaning one thing and writing another, but rather an explicit assertion that is false. Are there any such examples in the work of Gauss, I wonder? R. M. Robinson has called my attention to a lapse by Minkowski in which he asserts that the difference set of a tetrahedron is an octahedron. In his 1906 paper "Dichteste gitterförmige Lagerung kongruenter Körper," *Nachrichten der König. Gesellschaft der Wissenschaften zu Göttingen* 5(1904), 311–355, Minkowski writes

> For example, if K is a tetrahedron, then $1/2(K + K')$ becomes an octahedron with faces parallel to the faces of the tetrahedron.

Here K' is "the reflection of K with respect to the point O."

It's a curious sort of error since the polyhedron in question clearly has 12 rather than 6 vertices, namely, all pairs $a_i - a_j$, $i \neq j$, where the a_i are the vertices of the tetrahedron. The correct polyhedron is in fact the convex hull of the midpoints of the 12 edges of a cube, so it has eight triangular and six square faces.

TWO CONTRIBUTIONS FROM LEE SALLOWS

First the following:

"This computer-generated sentence contains two hundred forty-seven letters: four a's, one b, four c's, five d's, forty-four e's, nine f's, three g's, seven h's, eleven i's, one j, one k, three l's, two m's, twenty-nine n's, nineteen o's, two p's, one q, fourteen r's, thirty-one s's, twenty-five t's, seven u's, eight v's, seven w's, two x's, six y's, and one z."

Now that you've had a chance to verify the correctness of the sentence you may wonder how the computer generated it. Here is Sallow's description of how it's done.

"The algorithm that generated the above sentence implements an iterated function. Starting with a similar text, but using randomly selected totals, its true letter frequencies are determined and then substituted for these, the new version then furnishing the argument for the next iteration, and so on. The result is a series of approximations tending toward the goal. I like to picture this process as a machine that takes sentences as input and yields sentences as output, the latter coupled back to the input via a feedback loop. This makes it easier to see that a self-descriptive sentence is effectively a virus able to subvert the machine so as to get itself perpetually reproduced. Such a sentence has only to appear once at the input in order to trigger a closed loop of period 1 and thus be regurgitated ad nauseam (if you see what I mean). The only trouble is, there are still other viruses that will probably infect the machine first! These are the sentence chains of longer period, in any one of which it may easily become ensnared, and thus be prevented from converging onto a self-descriptor. How can we immunize the machine against such interloopers?

My answer is a modified machine that will scramble possible cycles through performing non-repetitively: Instead of correcting every total on every pass, I have it correct a single total chosen at random each time. Now no recurrent cycle can survive such irregular exchanges, except of course in the special case where the totals remain unchanged because already correct: a self-descriptor! [Note here the complete analogy with a neural network settling into a stable solution state, while avoiding latch-up in pseudo-solution states through 'jiggling.'] In fact the 'random' selection need not be truly random, provided only that the repetition period of its own pattern be longer than that of any possible loop the machine may fall into. Hence, any conventional pseudo-random number generator serves well. A few million iterations (mutations) normally suffice to evolve (naturally select) a viable solution (virus) provided one exists. If not we can try again with a modified text."

Sallows' second contribution involves neither letters nor numbers and is presented below without comment.

chapter 3

Historic Conjectures: More Sequence Mysteries

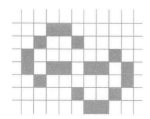

CONJECTURES

Among notable recent achievements in mathematics are the resolutions of some celebrated long-standing conjectures: the proof by Deligne of the Weil conjectures; by de Branges of the Bieberbach conjecture; and by Faltings of the Mordell conjecture. On the other hand, the Fermat problem,* the Riemann hypothesis, and the so-called Poincaré conjecture remain unresolved, though for brief periods it was claimed that they too had been settled. In any case, it seems timely to present some scattered facts about the origins of some of these problems. For most of what follows in this section, aside from the next paragraph, I am indebted to Professor M. R. Choudhury of the University of Dhaka, Bangladesh.

The statement commonly referred to as the Poincaré conjecture is the assertion that the only simply connected compact 3-manifold is the 3-sphere. However, as has been pointed out (see e.g., S. Smale, *Mathematical Intelligencer,* Vol. 12, No. 2, 1990) Poincaré never put this forward as a

---

*Now no longer a conjecture!

conjecture. He writes [Poincaré, H., Cinquième complément à l'analysis situs, *Œuvres VI*, Gauthier-Villars, Paris (1953), p. 498]:

> Il resterait une question à traiter: Est-il possible que le groupe fonda-mental de V se réduise à la substitution identique, et que pourtant V ne soit pas simplement connexe?
>
> There remains a question to be treated. Is it possible that the fun-damental group of V reduces to the identity even though V is not sim-ply connected?

There follows a paragraph in which the question is rephrased in terms of some of the concepts introduced in the paper, and then there is a final one-line paragraph: "Mais cette question nous entranerait trop loin," which roughly translated says, "But that question would carry us too far afield." Thus the question is presented neither as a conjecture nor as an open problem. Poincaré does not even guess at the answer.

By contrast, Weil is absolutely explicit in believing that his unproved results are true. He writes [Numbers of solutions of equations over finite fields, *Bulletin of the American Mathematical Society* 55(1949), p. 498],

> This, and other examples which we cannot discuss here, seem to lend some support to the following conjectural statements, which are known to be true for curves, but which I have not so far been able to prove for varieties of higher dimension.

As we now know from the work of Deligne, Weil's conjectures turned out to be cor-rect. It is interesting then to read what Weil has written on the subject of conjectures in general [Two lectures on number theory, past and present, *L'Enseignement mathématique* (2), 20 (1974), 87–110]:

> Here I may point out that in the old days, when we used the word "hypothesis" or "conjecture" (in German, *Vermutung*), this was not to be taken as simply a form of wishful thinking. Nowadays the two are often confused. For instance, the so-called "Mordell conjecture" on Diophantine equations says that a curve of genus at least two with rational coefficients has at most finitely many rational points. It would be nice if this were so, and I would rather bet for it than against it. But it is no more than wishful thinking because there is not a shred of evi-dence for it, and also none against it.

So Poincaré clearly did not make his conjecture, Weil definitely did. We leave it to the reader to decide whether Mordell's statement [On the rational solutions of the indeter-minate equations of the third and fourth degrees, *Proceedings Cambridge Philosophical Society* 21 (1922/23), 191–192] is a conjecture or wishful thinking:

> In conclusion, I might note that the preceding work suggests to me the truth of the following statements concerning indeterminate equa-tions, none of which, however, I can prove. . . .

(3) The equation

$$ax^6 + bx^5y + \cdots + fxy^5 + gy^6 = z^2$$

can be satisfied by only a finite number of rational values of x and y, with the obvious extension to equations of higher degree.

(4) The same theorem holds for the equation

$$ax^4 + by^4 + cz^4 + 2fy^2z^2 + 2gz^2x^2 = 2hx^2y^2 = 0.$$

(5) The same theorem holds for any homogeneous equation of genus greater than unity, say $f(x, y, z) = 0$.

MORE MYSTERIES

Many mathematicians have observed that the main impact of computers on mathematics has been to raise new problems rather than solve old ones. I will suggest that some of these new problems, though easy to formulate, may in fact be impossible to solve.

Among problems that would probably never have been posed but for the existence of computers, one of the simplest and probably the best known is the so-called $(3n + 1)$ conjecture due to Lothar Collatz. Let f be the function on the natural numbers $\mathbf{N}$ where

$$f(n) = \begin{cases} n/2 & \text{for } n \text{ even,} \\ 3n + 1 & \text{for } n \text{ odd.} \end{cases}$$

The conjecture is that for any n there is some k such that $f^k(n) = 1$, or in the language of dynamical systems, the orbit of every n contains 1. This has been verified for all $n < 10^9$.

A general class of questions of which this is a special case is easily described. For any number k and numbers a_i, b_i in $\mathbf{N}$, $0 \le i < k$, let f from $\mathbf{N}$ to $\mathbf{N}$ be defined by

$$f(kn + i) = a_in + b_i.$$

One can then ask questions about the orbits of points under f. For example, do they all contain the number 1? The Collatz problem corresponds to the case $k = 2$, $a_0 = 1$, $b_0 = 0$, $a_1 = 6$, $b_1 = 4$.

The main result on the general problem is due to John H. Conway [Unpredictable Iterations, *Proceedings Number Theory Conference, Boulder,* 1972] and asserts that it is undecidable. More precisely, Conway shows, even for the case when all the b_i are zero, that there is no algorithm for deciding whether the orbit of a given n contains the number 1. Of course this says nothing about the decidability of the Collatz problem, but it does show that there exist specific problems with numbers k and a_i that could in principle be calculated but for which the problem is undecidable. Further, Conway has come up with some interesting examples, referred to by Richard Guy as permutation sequences. The simplest of these appears in Guy's *Unsolved Problems in Number Theory* [Springer-Verlag, New York, 1981] and is given by the mapping T defined by

$$T(2n) = 3n,$$
$$T(4n + 1) = 3n + 1,$$
$$T(4n - 1) = 3n - 1.$$

T is easily seen to be a bijection, so all orbits are either cycles or bi-infinite sequences that approach infinity in both directions. One finds easily the cycles (1), (2, 3), (4, 6, 9, 7, 5), and (44, 66, 99, 74, 111, 83, 62, 93, 70, 105, 75, 59). The smallest missing number among these is 8, whose orbit starts out 73, 55, 41, 31, 23, 13, 10, 15, 11, 8, 12, 18, 27, 20, 30, 45, 34, 51, 38, 57, 43, 32, 48, 72, ... The orbit shows no signs of cycling after several thousand iterations in both directions although there are a number of "near misses." Note the 73 and 72 at the beginning and end of the sequence above. This happens again with 153 and 154, 161 and 162, 500 and 501, 790 and 791.

Question 1: Are there any other finite orbits under T?

Question 2: Is the orbit of 8 finite?

A striking feature of this last question is that it concerns a single sequence, as contrasted with the Collatz problem, which asks about the behavior of an infinite number of sequences. Of course it is meaningless to say that this question is undecidable, but later I argue that it may well be "unprovable"; that is to say, it might be that the orbit of 8 is in fact infinite, but that there is no proof of this from our usual system of axioms.

Further experimentation leads to further speculation. The smallest number not in any of the preceding orbits is 14, whose orbit appears to be heading resolutely for infinity at both ends. The next missing number is 40, and so on. Using *Mathematica* we found that all numbers up to 1000 belong to 54 disjoint orbits. We call the smallest number in an orbit the seed s. The elements of $T^n(s)$ are called *forward numbers* for n positive, and *backward numbers* for n negative. Of course, just as we have no proof that these orbits are not parts of cycles, we also don't know that they are disjoint. It is conceivable, for example, that some forward iterate $T^n(8)$ might eventually hit a backward iterate of $T^m(14)$. A crude statistical study indicates that the forward iterates contain roughly the same number of even and odd numbers. A consequence of this is that roughly half of the forward numbers are divisible by 3, since from the definition of T, a number is divisible by 3 if and only if its predecessor is even. Among the backward numbers, on the other hand, the odd numbers seem to outnumber the evens by about 2 to 1. This must clearly be the case, for an even backward number follows a down-jump of 2/3, while an odd backward number follows an up-jump of only 4/3. So there must be many more odds than evens, since the numbers $T^{-n}(s)$ must remain positive.

With the data at hand, one can prove that if there are any cycles other than those listed, they must have length at least 360. Further, one can intuitively argue that the existence of cycles becomes very "improbable." First note that any odd number gives a forward down-jump of (approximately) 3/4, while an even number gives an up-jump of 3/2. So if there are m odd and n even numbers, then for a cycle we must have $(3/4)^m(3/2)^n \approx 1$, and if the lowest number in the cycle, the seed, is around 1000, then this approximation must be close, that is, the ratio m/n must approximate 0.70951 to four decimal places. If the seed is greater than 10,000, then the smallest cycle would have length 665 and could occur only for $m = 276$ and $n = 389$. What are the "chances" of this happening?

What then is the moral of this story? We are all aware from the work of Gödel that no matter what system of axioms we work with, as long as just a bit of number theory is available, there are true propositions for which there is no proof. Yet we continue as diligently as ever in looking for proofs, and frequently we find them, mainly, I think, because of the problems we choose to attack. But problems like the one discussed here seem to be of a special sort. It seems to me overwhelmingly "likely" that the orbit of 8 is infinite, and it is correspondingly "unlikely" that there is a proof of this fact. Indeed, why should there be?

WHITEHEAD WIT

The topologist J. H. C. Whitehead was often asked for his views on the work of his uncle, the renowned philosopher Alfred North Whitehead. Eventually he developed a stock answer. When asked, "What do you think of your uncle's philosophy?" he would reply, "I really haven't thought much about it—but what do you think of *your* uncle's philosophy?"

Privacy-Preserving Protocols

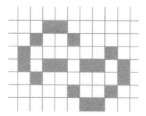

UNCONDITIONALLY SECURE PROTOCOLS

One of the exciting mathematical developments of the past decade was the discovery of so-called uncrackable public key codes. These are codes with the characteristic that everyone knows the method of encryption, but the amount of calculation required for an outsider to break the code is considered beyond present computational capabilities. In the best-known example, breaking the code was equivalent to finding the factors of, say, a 100-digit number, which was believed to be computationally infeasible.

A more recent but less well known development with somewhat the same flavor involves methods of conveying certain information that depends on other information that must remain secret. In these cases, however, it is literally impossible for anyone but a mind reader to learn the secrets. Here is a simple example. Some people, $P_1, \ldots, P_n$, say the members of a mathematics department, are interested in learning their average salary, but they are not willing to reveal their own salaries to anyone else. How can this be done? I put the problem to some of my colleagues, and they were not able to come up with an answer. I also mentioned it at a social gathering, and rather quickly a young woman who hadn't had a mathematics course since high school (and claimed she'd failed 9th grade algebra) proposed the following simple solution: P_1 chooses an arbitrary number x and tells it to P_2, who adds his salary and tells the total to P_3, who adds her salary and tells it to P_4, and so on, until P_n adds his salary and tells it to P_1 who adds her salary, subtracts x, divides by n, and announces the result. Clearly, no one has learned anything about anyone else's salary except what can be inferred from knowing the average.

Now, while it is true that in the above scheme no person acting alone can discover anything about the other people's salaries, the situation changes if people are allowed to collude. For example, if P_1 reveals x to P_3, then P_3 will know P_2's salary. Thus the scheme, or *protocol*, as it is called, is said to be *1-private* but not *2-private*. One may then ask if there are any 2-private protocols for this problem. The answer is that in fact, there is an *n*-private protocol, which is also easy to describe. A protocol is called *n-private* if no proper subset of the n people by colluding can learn anything about the complementary set except what can be inferred from their knowledge of the average. Here is how it works. Let s_i be the salary of P_i. Each P_i chooses n numbers s_{ij} subject only to the condition that they sum to s_i. P_i then writes each number S_{ij} on a separate piece of paper and hands the paper containing s_{ij} to P_j. Now each P_j announces the sum t_j of the numbers in his hand. The sum of the t_j is of course the sum of the s_i, as desired. The situation is represented by the matrix $S = (s_{ij})$ where the ith row sum is s_i and the jth column sum is t_j.

	t_1	t_2		t_k		t_n
s_1	s_{11}	s_{12}		s_{1k}		s_{1n}
s_2						
s_k	s_{k1}	s_{k2}		s_{kk}		s_{kn}
					S_k	
s_n	s_{n1}	s_{n2}				s_{nn}

To see that the protocol is *n*-private, suppose, say, the first k players collude. Then they will know all entries of S except those in the lower $(n - k) \times (n - k)$ submatrix S_k, and they will also know $t_{k+1}, \ldots, t_n$, so they will know the column sums of S_k. But if one knows only the column sums c_j of a matrix, then the row sums r_i can be any numbers subject only to the condition $\Sigma r_i = \Sigma c_j$. So the colluding players will know only the sum of the other players' salaries, which they would know anyway from knowing the sum of all the salaries.

The sum protocol can be used to learn other things, for example, the distribution of salaries, that is, the number of people at each salary level, without revealing who they are. To find out how many people have salary x, do the sum protocol where P_i's secret number is 1 or 0 according to whether his salary is or is not equal to x, and repeat the protocol for all values of x. A more efficient method is for P_i to choose the secret number $(n + 1)^{s_i}$. When the sum is computed, it is expressed in base $(n + 1)$, and the coefficient of $(n + 1)^x$ will be the number of people whose salary is x. The same trick can be used for a secret ballot. Suppose that the candidates for some office are labeled 1 through m. A person who wants to vote for candidate k should use the secret number $(n + 1)^k$. The sum protocol then gives the vote count.

What about other functions? For example, the maximum rather than the sum of the salaries? If an upper bound $\bar{s}$ of the salaries is known, the following procedure suggests itself. Use the sum protocol to find out how many people have salary $\bar{s}$. If the answer is zero try again with $\bar{s} - 1$, and so on until the sum is positive. The trouble is that one learns too much. One learns the maximum salary and also the number of people who earn the maximum. Is it possible to learn the maximum and nothing more? In the same spirit, is it possible to learn only the winning candidate(s) in an election but nothing about the distribution of votes? And a simple arithmetical question: the sum of n numbers can be computed n-privately—What about the product?

The answer to all of these questions is the same and is quite surprising. There exist t-private protocols for all of them if and only if t is less than $n/2$. Such protocols might be called *minority-private*. The existence of minority-private protocols was proved by Ben-Or, Goldwasser, and Wigderson [2] and independently by Chaum, Crépeau, and Damgard [3]. Given secret numbers $s_1, \ldots, s_n$ that may take on some finite set of values, any function of the s_i can be computed minority-privately. It suffices to consider functions on a sufficiently large finite field. A minority-private protocol is given for multiplication, which is somewhat more complicated than that for addition. (Everything we have described up to now could probably be understood by a competent 7th grader. The multiplication protocol is about at the level of an undergraduate abstract algebra course.) Once one has addition and multiplication, one has polynomials and hence all functions on a finite field. By suitable encoding most problems of the sort one is interested in can be transformed into a problem of calculating a function from integers to integers, although this is not immediately obvious, for example, in the secret-ballot problem, where one wants to know only which candidate won the election.

Perhaps even more striking than the sufficiency is the necessity of the condition $t < n/2$. This means there is no protocol, for example, for computing the product of n secret numbers that can maintain secrecy if half or more of the participants decide to collude. In fact, essentially the only functions that can be computed *majority-privately* are functions obtained using only the sum protocol. This was first shown by Chor and Kushilevitz [4] for Boolean functions and then by Beaver [1] for general integer-valued functions. Notice that we have nowhere up to now said what a protocol actually is, but have simply exhibited examples. This is fine as long as one is proving existence theorems. Analogously, to show that there is a "formula" for the roots of third and fourth degree polynomials, one simply displays them and checks that they work. On the other hand, to show *nonexistence* of such expressions for higher-degree polynomials, a strict formalization of the problem is necessary. Similarly, to prove nonexistence of majority-private protocols, one must have precise definitions of protocols and privacy and then develop the necessary theory to deal with these concepts; and the arguments are considerably more involved than those for existence.

As a special case of Beaver's result, we see that when there are only two people, essentially nothing can be learned privately, for example, whether they have the same secret number. On the other hand, from the existence theorem, we know that if a third party P_3 enters the picture and is able to give and receive messages, then P_1 and P_2 can learn whether or not they have the same number 1-privately, and P_3 will not know whether the answer is yes or no.

There is a good deal more to the theory than has been mentioned. For example, if one does not require unconditional security but only "uncrackability" in the sense described in the first paragraph, then it has been shown that any function can be computed n-privately, including the situation where there are only two people. In the so-called "millionaires problem" of Yao [5], for example, P_1 and P_2 can learn which of them has the larger salary and nothing else.

To conclude, let me return to the 7th grade level and describe a 1-private protocol that computes the maximum salary. For this we bring in an outsider P_0 who chooses some secret number x_0. The rules are then the following: if P_i's salary is $\bar{s}$ (the upper bound), she chooses some arbitrary positive number x_i. If not, her secret number is 0. Now do the sum protocol. If the sum is not x_0, then P_0 announces that $\bar{s}$ is the maximum. If the sum is x_0, play again, replacing $\bar{s}$ by $\bar{s} - 1$, and so on until the maximum is found. Notice that it is necessary to bring in P_0 because if the others played the game without him and at some stage the sum turned out to be x_i, then P_i would know that she was the only one getting the maximum. Similarly, the protocol with P_0 is only 1-private, because if P_0 gets together with a person earning the maximum salary, then the two of them will know whether or not anyone else is also earning the maximum.

I want to express my thanks to Donald Beaver of AT&T for much of the material I have presented and to Michael Hirsch of UC Berkeley for bringing this interesting subject to my attention. It seems there are more kinds of mathematics in heaven and on earth than are dreamed of in all your volumes of Bourbaki.

REFERENCES

1. D. Beaver, Perfect privacy for two-party protocols, *Harvard Tech. Report* TR-11-89, Aiken Computer Laboratory.
2. M. Ben-Or, S. Goldwasser and A. Wigderson, Completeness theorems for non-cryptographic fault tolerant distributed computations, *Proceedings 20th STOC,* 1988, 1–10.
3. D. Chaum, C. Crépeau, and I. Damgard, Multiparty unconditionally secure protocols, *Proceedings 20th STOC,* 1988, 11–19.
4. B. Chor and E. Kushilevitz, A zero–one law for Boolean privacy, *Proceedings 21st STOC,* 1989, 62–72.
5. A. C. Yao, Protocols for secure computation, *Proceedings FOCS,* 1982, 160–164.

SOMOS SEQUENCE UPDATE

In Chapter 1 we described some sequences defined by a simple recursion that for unexplained reasons always seem to yield integer terms. The sequences were originally introduced by Michael Somos and can be described as follows: Given an integer $k \geq 4$, a *Somos* (k) *sequence* is characterized by the recursion

(1) $$a_n a_{n-k} = x_1 a_{n-1} a_{n-k+1} + x_2 a_{n-2} a_{n-k+2} + \cdots + x_r a_{n-r} a_{n-k+r},$$

where $r = [k/2]$ and the x_i are given integers.

Since in (1) a_n is defined in terms of the preceding k terms, one must choose *initial values* for $a_0, a_1, \ldots, a_{k-1}$.

When I refer simply to Somos (k), I mean a k-sequence in which all the x_i and initial a_i are unity. It was first observed numerically and later proved that Somos 4, 5, 6, 7 always have integer terms, whereas Somos 8, 9, and presumably all the rest do not. But there still remains a doubly infinite family of Somos sequences that appear to have integer terms, although this has not been proved.

Motivated by the Somos phenomena, Dana Scott discovered that sequences with initial values unity and the following recursions have integer terms:

(2) $$a_n a_{n-k} = a_{n-1}^2 + a_{n-2}^2 + \cdots + a_{n-k+1}^2,$$

(3) $$a_n a_{n-k} = a_{n-1} a_{n-2} + a_{n-3} a_{n-4} + \cdots + a_{n-k+2} a_{n-k+1},$$

for k odd.

The integer property also holds for

(4) $$a_n a_{n-k} = a_{n-1} a_{n-2} + a_{n-2} a_{n-3} + \cdots + a_{n-k+2} a_{n-k+1}.$$

Proofs of integrality of (2), (3), and (4) have now been found by Raphael Robinson with an assist at one point from Dean Hickerson. The proofs are quite elementary, so we will present the one for (2). The arguments for (3) and (4) are similar. They involve finding rational functions that are *invariant* under the recursions. For (2) we define a new sequence (b_n) for $n \geq k$ by

(5) $$b_n = \frac{a_n + a_{n-k}}{a_{n-1} a_{n-2} \cdots a_{n-k+1}},$$

and the claim is that the b_n are constant, that is, $b_{n+1} = b_n$. To see this, note that

(6) $$a_n(a_n + a_{n-k}) = (a_{n+1} + a_{n-k+1}) a_{n-k+1}$$

because from (2) both sides of (6) are equal to $a_n^2 + a_{n-1}^2 + \cdots + a_{n-k+1}^2$. Dividing both sides by $a_n a_{n-1} \cdots a_{n-k+1}$ gives $b_n = b_{n+1}$. But from the initial conditions $b_k = k$. Hence $b_n = k$, so from (5),

(7) $$a_n = k(a_{n-1} a_{n-2} \cdots a_{n-k+1}) - a_{n-k},$$

which gives a new recursion for the a_n where the right-hand side is a polynomial (rather than a rational function) in the a_i. So integrality follows, as does the fact that the sequence reduced mod m is *periodic* for any m.

But although some problems have now been solved, further numerical explorations by Robinson brought to light a host of new structural properties of Somos sequences, some of them number-theoretic, others analytic. Since these results will appear elsewhere, I will just mention a few of them. First, all Somos sequences that give integers are periodic when reduced mod m for any m. Robinson proved this for Somos 4 and 5 but not for 6 and 7. For 4 and 5 the period as a function of m is unpredictable, but the following striking relation was observed: For all m except 2, the period mod m^k is equal to m^{k-1} times the period mod m. For 2 a somewhat more complicated pattern holds. Robinson also investigated which primes divide the various terms of the sequence and found that for Somos 4 and 5 (but not for 6 and 7), the terms divisible by a given prime were equally spaced. Thus in Somos (4), every fifth term is even, every seventeenth term is divisible by 11, and none is divisible by 5, while in Somos (5), every tenth term is divisible by 5, but none is divisible by 7.

In a different direction, Robinson investigated analytic properties of the sequences of arbitrary k and initial values and found in all cases tested that there were (unique) constants C and D such that

$$(8) \qquad\qquad a_n = C^{(n-D)^2} \phi(n),$$

where $\phi(n)$ has an oscillation with a well-defined period. The constants C and D depend on the initial values and on k but in an apparently unpredictable manner. Learning of Robinson's data, Clifford Gardner succeeded in finding explicit formulas for Somos 4 and 5 in terms of Jacobi elliptic functions. So in some sense the problem is starting to come full circle, since Somos originally discovered his sequences while studying properties of elliptic functions and became aware of some of the phenomena described here.

To a nonexpert the analytic and number-theoretic properties of the Somos sequences seem unrelated, but perhaps algebraic number theorists, who are accustomed to such things, will be able to make a connection. In any case, it is intriguing to see more and more properties of these sequences revealed by numerical exploration.

A TRUE STORY

Once upon a time there was a little girl named Clara who was barely three years old and had just learned how to count. She could tell how many chairs there were in the living room and the number of steps down from the front porch. One day her father decided to test her. "Look," he said, "I've brought you these four lollipops," but he handed her only three. Clara took the lollipops and dutifully counted, "One, two, four." Then she looked up a bit puzzled and asked, "Where's the third one?"

chapter 5

Surprising Shuffles

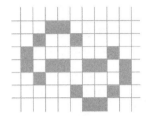

CAREFUL CARD SHUFFLING AND CUTTING CAN CREATE CHAOS

Imagine (if you can) a countably infinite deck of cards. Each card is marked with a different natural number, and initially the cards are arranged in their natural order with card 1 on the top and placed face down on a (finite) table.

Definition *A perfect n-shuffle consists in picking up the top* n *cards of the deck and interlacing them with the next* n. *Thus, if one executes a 5-shuffle on the deck in its initial ordering, the resulting ordering will be*

$$6,1,7,2,8,3,9,4,10,5,11,12,13, \ldots .$$

Now consider executing a sequence of shuffles, first a 1-shuffle, then a 2-shuffle, then a 3-shuffle, etc.

Conjecture. In the course of this sequence of shuffles, every card will come to the top of the deck infinitely often.

This conjecture is a mild modification of a conjecture of Richard Guy, as will be explained later, so I will refer to it from now on as Guy's conjecture.

Here are the orderings given by the first eight shuffles:

0	1,2,3, . . .
1	2,1,3,4,5, . . .
2	3,2,4,1,5,6, . . .
3	1,3,5,2,6,4,7,8,9, . . .
4	6,1,4,3,7,5,8,2,9,10, . . .
5	5,6,8,1,2,4,9,3,10,7,11,12, . . .
6	9,5,3,6,10,8,7,1,11,2,12,4,13,14, . . .
7	1,9,11,5,2,3,12,6,4,10,13,8,14,7,15,16, . . .
8	4,1,10,9,13,11,8,5,14,2,7,3,15,12,16,6,17,18, . . .

Note that at this point cards 1 through 6 have made their way to the top of the deck, 1 having been there already three times. However, card 7 won't get there until shuffle 78, a first indication of the sort of erratic behavior to be described shortly.

Getting back to Guy's conjecture, what basis is there for believing it may be correct? There are two arguments for it, the first "empirical," the second "probabilistic" (quotation marks here mean that what we are about to say is, rigorously speaking, nonsense). We consider first the probabilistic argument suggested by Raphael Robinson. Note first that card c stays in its place until the $[c/2]$th shuffle, after which it may bounce around in a seemingly turbulent manner; but it can never reach a position further than $2n$ from the top of the deck on the nth shuffle, so the position of c on the nth shuffle is some number between 1 and $2n$. Now assuming (here's where the nonsense sets in) that the position of c is random, the probability that it is not at the top of the deck is $1 - 1/(2n)$, and (more nonsense) assuming that successive positions are independent, we see that the probability that c never reaches the top of the deck is

$$\prod_{n=1}^{\infty}\left(1 - \frac{1}{2n}\right)$$

which is zero because of the divergence of the harmonic series.

As for "empirical evidence," Ilan Adler has checked all cards from 1 to 5000, listing for each card the number of the shuffle at which it reaches the top of the deck, and the results are interesting. Things go along comparatively quietly until we get to card 39, which takes 13,932 shuffles to reach the top, after which things settle down again, although card 43 requires 30,452 shuffles. But then there is a major explosion. Card 53 takes a mere 30 shuffles, but then card 54 goes on a wild rampage (card 54, where are you?), finally making it to the top on shuffle 252,992,198.

Collecting all the data took about 80 hours of computer time on a NeXT work station, but most of that time was taken up with three "monsters," 4546, 3729, and the current world champion, 3464, which took respectively, 2,263,846,432, 15,009,146,841, and 21,879,255,397 shuffles. Of course, this sort of behavior is what one would expect from the fallacious probability argument. The longer a card stays away from the top, the longer it is likely to continue to do so; i.e, the chance of "choosing" a 1 on the millionth shuffle is one in a million.

Since perfect shuffles behave so wildly, perhaps one should look at something simpler. Instead of shuffling, one might try simply cutting the cards. The traditional cut takes the top, say, n cards and places them on the bottom of the deck. In our model, however, the bottom is too far away, so instead let us define an *n-cut* to interchange the top n cards with the next n. Thus a 5-cut on the original order produces

$$6,7,8,9,10,1,2,3,4,5,11,12,13, \ldots$$

If we now perform n-cuts in consecutive order starting with a 1-cut, then a 2-cut, etc., it is trivial to show that Guy's conjecture is true, for card c remains in place until the $[c/2]$th cut, after which it moves up the deck by one on every other cut until it hits the top, whereupon it jumps down and then again proceeds to work its way back up to the top. So to make things interesting, instead of merely cutting, we will cut, but after each cut we discard the top card. The question is then whether every card is eventually discarded. The statistical behavior here is somewhat more restrained than for the shuffles, although there are occasional spurts. For example, card 752 survives over nineteen million cuts before being discarded. It seems, though, that this problem may be tractable. The orbit of a given card has a clear pattern, which the reader will easily find by working a few examples, and the question of how long a card will survive boils down to a question in number theory—which, however, as of this writing has not been settled.

A little bit about the origin of these questions: It all began with a problem proposed by Clark Kimberling that appeared in *Crux Mathematicorum* volume 7, number 2 (Feb. 1991). Kimberling considers the following array:

$$
\begin{array}{cccccccccc}
\underline{1} & 2 & 3 & 4 & 5 & 6 & 7 & 8 & 9 & 10 & \ldots \\
2 & \underline{3} & 4 & 5 & 6 & 7 & 8 & 9 & 10 & 11 & \ldots \\
4 & 2 & \underline{5} & 6 & 7 & 8 & 9 & 10 & 11 & 12 & \ldots \\
6 & 2 & 7 & \underline{4} & 8 & 9 & 10 & 11 & 12 & 13 & \ldots \\
8 & 7 & 9 & 2 & \underline{10} & 6 & 11 & 12 & 13 & 14 & \ldots \\
6 & 2 & 11 & 9 & 12 & \underline{7} & 13 & 8 & 14 & 15 & \ldots
\end{array}
$$

Here each row is obtained from the previous one by a sort of leapfrog procedure. Start with the number to the right of the diagonal term, which is underlined. Then go to the number to the left of the diagonal, then back to the second number to the right, then the second number to the left, etc., until you reach the first number in the row. Then jump back to the right and leave the remaining numbers in their natural order. Once a number appears on the diagonal, it is expelled. Kimberling now asks, "(a) Is 2 eventually expelled? (b) Is every number eventually expelled?" The procedure is easily interpreted as a shuffle, which I will call the Kimberling Shuffle (sounds like the name of a nineteen-thirties dance craze), in which one discards card n on the nth round, then reverses the order of the first $(n-1)$ cards and interlaces them with the next $(n-1)$.

Richard Guy noticed right away that the answer to (a) was yes, and in fact 2 is expelled on row (shuffle) 25, as a fairly easy hand calculation shows. Guy then conjectured that (b) is also true, and with the help of his grandson Andy Guy, who is studying computer science at Cambridge, verified the conjecture for all numbers up to 1200. Their table shows the same sort of wild behavior as the one for the perfect n-shuffles. In a private communication Guy has written, "I'd guess all numbers are expelled, but I also guess that no one's going to prove it." So he has actually made two conjectures, with the interesting property that if either one is confirmed the other one won't be.

A SPANISH SELF-DESCRIPTOR

In Chapter I Lee Sallows gave a recipe for constructing self-descriptive sentences. Obviously the procedure is language-independent, so here is a Spanish version constructed by Miguel A. Lerma of the computer science department of the Universidad Politecnica of Madrid.

> ESTA FRASE CONTIENE EXACTAMENTE DOSCIENTAS TREIN-
> TA Y CINCO LETRAS: VEINTE A'S, UNA B, DIECISEIS C'S, TRECE
> D'S, TREINTA E'S, DOS F'S, UNA G, UNA H, DIECINUEVE I'S,
> UNA J, UNA K, DOS L'S, DOS M'S, VEINTIDOS N'S, CATORCE O'S,
> UNA P, UNA Q, DIEZ R'S, TREINTA Y TRES S'S, DIECINUEVE T'S,
> DOCE U'S, CINCO V'S, UNA W, DOS X'S, CUATRO Y'S, Y DOS Z'S.

I understand Sallows also has a Dutch example.

A RE-VIEW OF SOME REVIEWS

Among his many noteworthy accomplishments, Paul Erdös may well hold the all-time world record for the number of papers which he has co-authored. It is interesting, therefore, to note that there is at least one paper of which he is the sole author which was, nevertheless, in some sense a collaboration. Irving Kaplansky had just finished writing a review of a (joint) paper by Erdös when he encountered the author himself and mentioned that he had admired the result but wondered whether the proof of the main theorem, which ran over a page and a half, couldn't be substantially shortened. Erdös took another look and quickly found that indeed it could. The excerpt from *Math. Reviews* [Vol. 7, 1946, page 164] reproduced below provides the full story.

REFERNCES

Anning, Norman H., and Erdös, Paul. Integral distances. Bull. Amer. Math. Soc. 51, 598–600 (1945). [MF 12821]

The authors show that for any n there exist noncollinear points $P_1, \ldots, P_n$ in the plane such that all distances P_iP_j are integers; but there does not exist an infinite set of noncollinear points with this property. [Cf. the following review.] *I. Kaplansky* (Chicago, Ill.).

Erdös, Paul. Integral distances. Bull. Amer. Math. Soc. 51, 996 (1945). [MF 14475]
The paper reads as follows.

"In a note under the same title [see the preceding review] it was shown that there does not exist in the plane an infinite set of noncollinear points with all mutual distances integral.

"It is possible to give a shorter proof of the following generalization: if A, B, C are three points not in line and $k = [\max (AB, BC)]$, then there are at most $4(k+1)^2$ points P such that $PA - PB$ and $PB - PC$ are integral. For $|PA-PB|$ is at most AB and therefore assumes one of the values $0, 1, \ldots, k$, that is, P lies on one of $k+1$ hyperbolas. Similarly P lies on one of the $k+1$ hyperbolas determined by B and C. These (distinct) hyperbolas intersect in at most $4(k + 1)^2$ points. An analogous theorem clearly holds for higher dimensions."   *I. Kaplansky* (Chicago, Ill.).

So here is a rare example of a paper published in its entirety in two different journals—and now this makes it three (perhaps another world record).

Hundreds of New Theorems in a Two-Thousand-Year-Old Subject: Where Will It End?

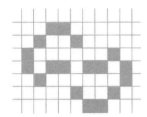

FROM EUCLID TO DESCARTES TO *MATHEMATICA* TO OBLIVION?

Once again our subject is the interaction of mathematics and computers.

Computers can solve mathematical problems. They can also pose them, and now, it seems, they may be capable of killing off whole branches of the subject, much in the same way professional chess may be killed off when and if a computer becomes the world champion. A potential victim is the most venerable branch of all, Euclidean geometry, meaning specifically, the geometry to be found in the books of Euclid.

These thoughts were stimulated by some recent explorations by Clark Kimberling involving centers of triangles, examples of centers being the centroid, the center of the inscribed or circumscribed circle, the intersection of the altitudes (orthocenter), etc.—where "etc." includes some 91 different notions of center described by Kimberling. It is easy to see how one can define centers essentially at will. Namely, if a, b, and c are the lengths of the sides of a triangle and $f(x, y, z)$ is some function that is symmetric in y and z, then the corresponding center is the point whose distances from sides a, b, and c are proportional to $f(a, b, c)$, $f(b, c, a)$, and $f(c, a, b)$, respectively. The function f is said to give the *trilinear coordinates* of the given center. Of course, interesting centers are those that correspond to natural geometric constructions. About half of the centers on Kimberling's list have appeared in the literature, and the rest were discovered (invented) by him. One of the striking things that shows up experimentally is that there are an enormous number of collinearities among the 91 centers. Indeed, all 91 points can be covered by only 103 different lines (out of a possible 4095 if no three of the points were collinear).

The classical example of collinearity is the Euler line, which passes through the centroid, the circumcenter, and the orthocenter, with the centroid lying two-thirds of the way from the orthocenter to the circumcenter. This result appears as an exercise in many books on elementary analytic geometry. It is also known that the Euler line passes through the center of the 9-point circle. If you don't remember the definition of the 9-point circle, well, neither do I, but in any case its center is midway between the circumcenter and the orthocenter. Kimberling finds eight other centers that also lie on the Euler line. It should be emphasized that these collinearities have all been discovered with a computer by taking a few numerical values for a, b, and c and noting that the various centers line up to within ten or so decimal places. Of course, all but a handful of these collinearities have never before been seen by man or beast. One might think of them as hundreds of new theorems in search of proofs.

To take one example, the *Fermat point* of an acute triangle is the point that minimizes the sum of its distances from the three vertices. The *Napoleon point* (said to have been discovered by the Emperor himself before he gave up geometry in favor of world conquest) is found by constructing equilateral triangles on each of the three sides of a triangle and then connecting their respective centers to the opposite vertices of the original triangle. The three lines are (theorem) concurrent, and their intersection is the Napoleon point.

Now it turns out that, for no apparent reason, the line through the points of Napoleon and Fermat passes through the circumcenter, a fact that one probably would not have predicted. However, once the phenomenon has been observed, a proof is immediately at hand. In principle, of course, this was already true after Descartes's discovery of coordinate geometry, but in practice it might be quite difficult and "messy" to derive and solve the polynomial equations involved. This is where symbol-manipulating programs such as *Mathematica* come in. *Mathematica* doesn't mind messy. Just feed it the trilinear coordinates of the three points, which in this case are

$$\text{Fermat: } \{\operatorname{cosec}(A + \pi/3), \operatorname{cosec}(B + \pi/3), \operatorname{cosec}(C + \pi/3)\},$$

$$\text{Napoleon: } \{\operatorname{cosec}(A + \pi/6), \operatorname{cosec}(B + \pi/6), \operatorname{cosec}(C + \pi/6)\},$$

$$\text{Circumcenter: } \{\cos A, \cos B, \cos C\},$$

Where A, B, C are the angles of the triangle. Multiply the first two rows by $\sin A \sin B \sin C$ to eliminate denominators. *Mathematica* then expands the determinant as a polynomial in sines and cosines, which in this case turns out to be zero, Q.E.D.

As another curiosity, there is a second Napoleon point, in which the construction is the same except that this time the equilateral triangles are constructed on the inside rather than the outside of the original triangle. In this case the line through this second Napoleon point and the Fermat point passes through the center of the 9-point circle rather than the circumcenter. How would one ever find a synthetic proof of that, I wonder—but of course *Mathematica* takes it right in stride.

Surely this is a rather strange state of affairs. Everything is being done by the computer. Program A goes on a voyage of exploration and comes up with a vast number of theorems. Then program B takes over and supplies the proofs, and, while all this is going on, the investigator just sits back and watches. The robots have taken over. It makes one reflect a bit on what we are trying to achieve in doing mathematics. It is certainly impressive to suddenly learn hundreds of new facts in a discipline that people have worked in for more than two millennia. But mathematics, and science generally, is concerned with much more than compiling a huge catalogue of facts. The hope is to find general principles from which the facts can be deduced, and the robots don't seem to be very helpful for this. They tell us what is true but don't tell us why. They supply lots of information but little insight.

TRIANGLES ANYONE?

If you ask someone to draw a triangle, they will almost always draw something like this or this

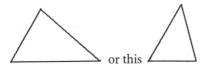

but almost never like this

Donald Newman has pointed out, however, that almost any way you look it, a "random triangle" is much more likely to be obtuse than acute. How does one pick a triangle at random? One possibility is to pick three positive numbers for the angles at random subject to the condition that their sum is π. If this is the criterion, then it is quite easy to see that three out of four triangles are obtuse. This is equivalent to the observation that the set of points in the unit 2-simplex in R^3 all of whose coordinates are less than $1/2$ has one-fourth the area of the simplex. Alternatively, one might pick the side lengths at random from the unit interval and ask, Of those triples that correspond to triangles, what fraction correspond to acute triangles? Here the analysis is more complicated, but the answer turns out to be $\pi/4$. Still another possibility is to pick three points at random on the unit circle. Here the answer is again that three times out of four the triangle will be obtuse. Reason: the triangle will be acute if and only if the origin lies in the convex hull of the three points, which is equivalent to requiring that the origin be a linear combination of the three points considered as vectors and that the coefficients all have the same sign; this will happen one time in four.

TRIANGLES AND TEACHING

At a very early age most children become familiar with geometric objects like triangles, squares, and circles. Here are a few suggestions, some of them based on personal experience, on how to take advantage of this familiarity to show them some genuine mathematics, as opposed to the arithmetic that has been taking up so much of their time. I recall my own first encounter with a mathematical proof. I was in fifth grade when a fellow student taught me about the sum of the angles of a triangle and showed why it was true. He didn't give the Euclidean proof (Book 1, Proposition 32), which, as you will recall, depends on knowing things about parallel lines being cut by transversals, but rather he presented the "moving stick" proof illustrated by the sequence below:

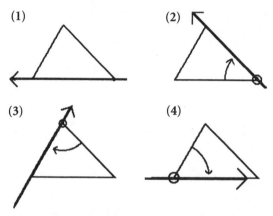

The stick is placed along the base of the triangle pointing west and is then pivoted successively about the right vertex, then the top vertex, and then the left vertex so it ends up back on the base but pointing east. Therefore, in fourth-grade parlance, it must have turned halfway around. I was charmed—as who wouldn't be, I thought in my naiveté. This was a high point in my mathematical experience until several years later when I learned about Proposition 47, the one about the square of the hypotenuse. Of course, I had no idea back then that one could devote one's life to finding and proving things like these (though not such nice things, of course) and even get paid for doing it (sometimes I still find this rather remarkable). But to get on with the story, a few years ago I thought about the moving-stick argument again, was I as interested in trying to convey to outsiders, children included, the sort of thing that goes on in mathematics, and I observed that obviously, if you take the stick around the triangle a second time, you will end up pointing west again. So I asked myself, and I now ask you, is the stick then in exactly its starting position, or has it perhaps been shifted to the right or left depending on the shape of the triangle? If you want to think about this you'd better stop reading, because I'm about to give the answer.

The stick is in its original position. Why? Well, the first trip around the triangle reversed the stick's orientation, and an orientation-reversing mapping has a fixed point. But the second trip was exactly like the first, so it must have the same fixed point and therefore there is no shift. To continue, just today as I started writing this up, I said to myself, o.k., there is a point on the moving stick that ends up back at its starting point. Which point is it? Let's make this a multiple-choice question. The fixed point is (a) the midpoint of the base, (b) the foot of the altitude, (c) the intersection of the base with the bisector of the opposite angle, (d) none of the above. I leave the answer to you with the following remark: this really *is* a quickie.

Getting back to the children, the moving-stick argument can of course be applied to other polygons. Of course, now it is not simply a question of whether or not the stick is reversed after its tour of the polygon, but also of how many times it turned, and this provides a nice intuitive introduction to the very basic notion of winding number. The fixed-point argument that we gave for the triangle works in fact for all odd-gons and fails in general for even-gons.

In another direction, once you have Proposition 32 and its corollary that an exterior angle of a triangle is the sum of the opposite interior angles, you can bring in circles and prove, for example, that the inscribed angle is half of the central angle (this also uses the fact that isosceles triangles have equal base angles (Euclid, Proposition 5), but I think most children would accept this as obvious).[1] If you've forgotten how the proof goes, the pictures below should jog your memory.

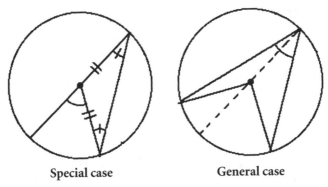

Special case   **General case**

Can average ten-year-olds be expected to follow all this? Perhaps not, but we won't know until we try. And how will we know whether they have really understood? One way is to ask them to do a similar problem on their own. For example, what is the relationship between angle *A* and arcs *a* and *b*?

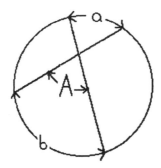

---

[1] Of course, we grown-ups know that even "obvious" propositions have to be proved, but surely the concept of rigor has no place in the fifth grade. In fact, the only reason for rigor at any level is to make sure we don't make mistakes. There is nothing sacred about rigor for its own sake. It was the failure to recognize this fact that led to the demise of the new math.

(There's no need to frighten the kids by using the customary Greek letters. I remember my own terror the first time I encountered a ξ.) A sequence of hints may be useful here. First, if they have difficulty, tell them what the answer is and ask them to prove it; if that doesn't work, tell them they must draw one more line; then tell them which line to draw, etc.

For those who are able to work it out, the next challenge might be this one:

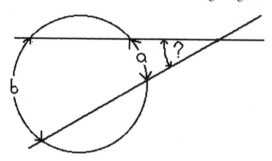

and so on.

With all the current concern about the "innumeracy" (dreadful expression!) of schoolchildren in the United States, perhaps a modest dose of triangles would be helpful in the enterprise of turning out students who are more "numerate." Actually, one of the nicest aspects of this project is that it has nothing at all to do with numbers.

chapter 7

Pop Math and Protocols

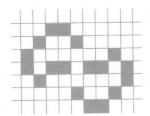

MORE ON PROTOCOLS: PLAYING GAMES OVER THE TELEPHONE

For almost all of what follows I am indebted to Imre Barany.

1. Coin-Tossing

A mathematician in Moscow and another in New York are talking long distance about a joint paper they have written which they have been invited to present at a meeting in Paris. Because both of them would like to go but only one can, they decide that the only fair thing is to leave it to chance. "All right," says the Russian, "I just tossed a coin. Is it heads or tails?" "Wait a minute," says the American, "that's not fair." "What's the matter?" says the Russian. "Don't you trust me? All right then, I give you the number 32357650571 08466121397469644230097884888530062931908450094100370 62 5655448971039595527235426169795539, and if you tell me correctly whether its largest prime factor is congruent to 1 or to −1 modulo 4, you can give the paper." After the American makes his guess the Russian reveals the prime factorization, which in this case happens to be 5612369956602102055876627916638107484790 3158831

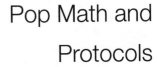

451 × 57654165390541998 80123699003158831450065809801648 9, and the American can verify the multiplication and the primality of the factors.

So until and unless someone finds a fast algorithm for factoring, this is a pretty good cheat-proof method to replace coin-tossing.

2. Scissors, Paper, Rock—and Two-Person Games in General

The above trick extends easily to playing any two-person game that does not involve chance moves. Suppose, for example, we want to play scissors, paper, rock. I will then send you a big number whose largest prime factor is congruent to 1, 2, or 3 modulo 5, based on whether I want to play scissors, paper, or rock. You then announce your choice directly, no encoding being necessary, after which I reveal the prime factorization of the transmitted number. The same thing will obviously work for games with n strategies by encoding them in numbers whose largest prime factors are congruent to $1, 2, \ldots, n$ for a suitable modulus. Thanks to Dirichlet, we know we will always be able to find the needed primes.

3. Games with Three or More Players

Interestingly, the protocols become much simpler and more secure when there are more than two players. In an n-person game, each player has a given finite set of *strategies*, and a *play* of the game consists in each of the players picking elements from these sets *simultaneously*. It is the simultaneity feature that presents the problem for implementing the game by long distance. The solution lies in the mechanism of strategy "sharing." Namely, if player i has m possible strategies, these are numbered 0 to $m - 1$, and this numbering is known to all of the players. If the player now decides to play strategy k, he chooses $(n - 1)$ numbers modulo m whose sum is k and distributes one of these numbers to each of the other players. After everyone has done this, the strategies are revealed, and, clearly, any player who lies about which strategy he has chosen will be caught by the others. Unlike the two-person situation, this protocol is "unconditionally" secure, meaning it does not depend on the assumption that certain calculations are computationally infeasible.

4. Bridge

Bridge and, indeed, all card games do not fall under example 3 because they involve choices by the players and also chance moves of Nature, namely, the deal of the cards. Nevertheless, there are nice randomizing protocols. This one for bridge is due to Dmitrij Grigoryev: East, West, and South agree on an encoding of the cards of the deck, say with the numbers from 1 to 52. North, not knowing the encoding, randomly assigns 13 numbers to each of the others, who will then know their hand. The unassigned numbers make up North's hand, but of course, he does not yet know what these numbers represent. No problem. To find out whether he has the ace of spades, he tosses a coin, and say it falls heads. He tells the result to West. If she does not have the ace, she tells heads to South, and if she does have the ace, she tells South tails. And in general, players either pass on or "reverse" the message they receive depending on whether or not they have the ace. Thus,

North has the ace if and only if East tells him heads. He then does the same thing for the two of spades. Proceeding through the deck in this manner gives North the needed information without revealing anything to the other players.

It is amusing to observe that it is possible to implement the procedure above while imposing an additional condition that players shall not be allowed to communicate directly with their partners. This requirement presents a problem because North must deal a hand to South. The problem is solved as follows: South invents a new encoding of the cards. Thus, 1 is zig, 2 is zag, etc. This new encoding is passed along to North using East as the intermediary. North then deals 13 of the "zigs, zags, ..." to South using West as the intermediary, and again no one but South gains any new information.

This example raises the question as to what sort of communication networks can be used for the card-dealing operation. The answer is that a necessary and sufficient condition is that the network be doubly connected; that is, there must be at least two disjoint communication routes between every pair of players. One sees how this property was used in the example above, in which two different routes had to be used from North to South. The necessity of the condition means that if the graph contains a vertex whose deletion separates the graph into two components, then a secret deal of the cards is not possible. We will return to this point later.

5. Poker

Note that the bridge-dealing protocol depends crucially on the fact that the entire deck is being dealt out. For games like poker, this is, of course, no longer true, so a different protocol must be devised. Here is Barany's description of the method of Barany and Furedi in the case of three players—an American, a Russian, and a Hungarian. The American has a Russian-Hungarian dictionary (giving the names of the cards), the Russian has an American-Hungarian dictionary, and the Hungarian knows all three languages. The American now deals the Russian 5 cards by sending her their Russian and Hungarian names, which, of course, are meaningless to the American. The Russian recognizes her 5 cards and now also knows their Hungarian names, so she then scratches out the words with these names in her American-Hungarian dictionary, after which she deals 5 of the remaining cards back to the American. Finally, either of the two may deal cards to the Hungarian. Of course, this will work for only one deal of the cards, because in the course of the deal the American and the Russian are starting to acquire a Hungarian vocabulary. For each deal, therefore, the Hungarian must make up a new pair of dictionaries.

In Chapter 4 we saw how protocols can be used not for playing games, but for gaining information without giving away secrets. A typical example is the case of people with secret numbers—their salaries or ages, for example—who are interested in learning only the sum of these numbers. As in example 3 above, the solution lies in sharing. Each player chooses $(n - 1)$ numbers whose sum is his salary and distributes them to the others. Then everyone announces the sum of the numbers they have received, and the sum of these numbers is, of course, the sum of the original secret numbers. This simple procedure has the further feature that it is "n-private," meaning that even if any proper subset of the people get together and pool their information, they will still be unable to find out anything about the secret numbers of the others (aside from what they would know anyway from their knowledge of the sum). However, the sum function is special in that it is, in a suitable sense, the only function that can be computed n-privately. Thus, for example, if you want to know the maximum rather than the sum of the salaries, the best you can achieve is "minority privacy," which means that there is a protocol such that no set of fewer than half of the participants can learn anything about the complementary set by pooling information. Indeed, there is what might be called the fundamental theorem of protocols, which asserts roughly that any function can be computed minority-privately, but only a few, like the sum, can be computed n-privately. Thus, for most functions there is no protocol that is guaranteed to preserve privacy if some group of half or more of the participants is interested in gaining additional information. We will illustrate this impossibility result for the special case of the unanimity problem. The argument, due to Barany, requires a somewhat greater degree of concentration, but I think that readers who are willing to make the mental effort will find it rewarding.

A motion brought before committee C will pass if and only if it is not vetoed by any of the members. The committee wishes to know whether or not the motion passes and nothing else. The assertion is that no matter what protocol is used, if partition of the set of voters is partitioned into sets A and B then members of one of these two sets will gain information. This shows, for example, that if there are 7 voters then there can be no 4-private protocol. The argument runs as follows: Consider the case where the motion fails and there are negative votes in both A and B and assume the protocol gives no additional information to A or B. Now imagine an outsider who knows the protocol and the fact that the motion has failed and is given a transcript of all the messages M_{AB} sent between members of A and members of B. Then it must be the case that the outsider cannot from this information eliminate the possibility that all members of A approved the motion; for if he could, then so could the members of B, because they have access to the same information. So they would be able to conclude that at least one member of A cast a veto, which they are not allowed to know. This means that there exists a sequence of messages M_{AA} between members of A that is consistent with the protocol in the case where all members of A favor the motion. Symmetrically, there exist messages M_{BB} that together with M_{AB} would be consistent with the protocol if all members of B favored the motion. But then M_{BB} is consistent with M_{AB} and M_{AA} because the messages M_{AA} cannot affect the legitimacy of the messages M_{BB}. This yields a contradiction because it shows that a set of messages that would occur when all voters favor the motion would, nevertheless, yield the conclusion that the motion has failed.

The reader who has persevered through the above reasoning should note that nowhere have we defined what is meant by a protocol or even a message, and that we have used only the property that the set of messages, whatever one may mean by this, is able to give exactly the desired information and nothing else.

Recall, finally, that in the problem of secretly distributing a deck of cards, we required that the communication network be doubly connected. This condition fits in with the impossibility result above. Namely, suppose removing some player P splits the remaining players into sets A and B whose only means of communication is through P. Now, the messages exchanged between A and B must determine uniquely how the cards not held by P are divided between A and B, but P has access to all of these messages and therefore, knowing the protocol, would know this distribution, which is not allowed. A corollary of this is that in games with only two players, there is no protocol for dealing secretly. (On the other hand, one of the earliest results in this area, by Adleman, Rivest, and Shamir, shows that if one assumes the existence of computationally hard problems, for example factoring, then the two-player case can also be handled.)

REFERENCES

I. Adleman, R. Rivest, and A. Shamir, "Mental poker," *The Mathematical Gardner,* London: Wadsworth International (1981).

Imre Barany and Zoltan Furedi, "Mental poker with three or more players," *Information and Control* 59(1983), 84 – 93.

POPULAR MATHEMATICS

How do you convince the man on the street about the Marvels of Mathematics? Two of the better-known attention-getters are the story of the emperor and the wheat and the one about the monkeys and the typewriters. The first story involves the grateful emperor who promised to reward his adviser by giving him grains of wheat on the squares of a checkerboard, one grain the first day on the first square and 2^{n-1} grains on the nth day on the nth square. The point was that this was more than the annual wheat production of the entire empire (to be exact, 18,446,744,073,709,551,615 grains, though this number was probably not given in the original tale). The mathematical moral of the story is, I suppose, that $2^{64} - 1$ is a bigger number than most emperors or men on the street realize.

As for the monkeys, there are usually six of them and they sit all day at their typewriter pecking away "at random." The assertion is that in time they will, "according to the laws of probability," type out all of Shakespeare's plays or, if you prefer, all the books in the British Museum in alphabetical order. Upon closer scrutiny this turns out to be a rather empty assertion because the laws of probability, quote unquote, turn out to mean exactly that the monkeys will do what they are supposed to do. Nevertheless, the claim is supposed to elicit oohs and ahs from the intended audience.

In this note I will suggest an illustration from the Wonderful World of Mathematics that includes both the powers of two and the British Museum and has the advantage that it is a precise, true statement (or, as we mathematicians say, a theorem), which nevertheless may seem to the uninitiated rather improbable. In fact, even the initiated may need a bit of convincing. The statement is that if the emperor had continued doubling the amounts of grain, there would eventually come a day when the leading (leftmost) zillion decimal digits of the number of grains would be exactly a digital encoding of the books in the British Museum. Furthermore, unlike the situation with the monkeys, one can pin down with certainty a definite date before which the desired number will have appeared. If we are a bit more modest in our demands and ask, say, for the day on which the number of grains added starts out with the digits 1992, we can exhibit the number explicitly. It happens to be 4077, because $2^{4077} =$ 19920137050879526265940717901546943332289 9675187565570904585549027243405908881656045689928423881065052202245085609 2954326059368861881134171738359219577127036017550974725913969809753229797 0527327428449632998646970057739036445181416639773308628957103251802045575 2952011244361970405804842213569598738857324612974050477756095477956591854 4583529293052281445175618539867910317288570881194541928629192648729206734 5606406185757137135380738704848884339598648927747075851697732623493514078

83993794451733264390721694007301018073452223453811908661381477128826783309006143922669341433058743019071586844344804050363054126636829830293768181717125429760654069664165918404279140428508802665119508323796752843572233612755191236217024322204874175245961379676637905113440547833845268843347369847734151454764458652171217413606969170060521343134040597782175502078927760751322716533966205843842981716595660622451346489263942349072378214190233740012032631697281797408521686200442571129300122590233653983306139144796219210553134483041738024837350879350484938262831338033830178875321688031458529831869663147033206367958245585806059674908652956376992134662982433124093647599427395439396364873350091299670503904558487495571467978179185913811628025131 8272.

For later reference, we note that the distribution of digits in the number above is quite uniform. There are only 118 zeros, and there are 137 threes, while the mean is 124, but these deviations from uniformity are well within the usual allowable limits. (Something that I have not been able to account for mathematically, however, is the fact that 4077 also turns out to be the last four digits of my social security number.)

In more formal terms, the theorem states that for any sequence of decimal digits N, there exists a natural number n such that the leading digits of 2^n are the sequence N. This result may even have commercial possibilities. Send us the sequence made up of the day, month, and year of your birth (along with d dollars) and you will receive by return mail your own personal n, guaranteed to last a lifetime.

Another nice feature of this theorem is that its proof depends only on knowing how to do long division, plus a mathematical fact that even an emperor should be able to understand, namely, the famous *Schubfachprinzip* associated with Dirichlet, which says that if you put a zillion and one objects in a zillion boxes, then one of the boxes must contain more than one object. (Emperor: You mean you have to go to graduate school in mathematics to learn *that*?) How does this apply in the present case? Well, start with the first n such that 2^n has at least $d + 1$ digits. Then, somewhere among the next $10^{d+1} + 1$ powers of 2 you will get two different numbers whose first $d + 1$ digits are the same. Now divide the larger by the smaller, and recall the rules of long division. The number you get will either (a) start with a one followed by at least $d + 1$ zeros (and then other stuff) or (b) start with $d + 1$ nines. (Whether you get a or b depends on the relative sizes of the first digit where the two numbers differ.) Let us denote the quotient in case (a) by Z. The claim is that if one takes successive powers of Z, the first d digits of the numbers obtained will run through all d-digit numbers in their natural order. This is because *if any number N is multiplied by Z, either the first d digits of the product will be the same as those of N or else the dth digit will increase by 1*. To see this, consider the number Z' with the same number of digits as Z whose leading digit is a 1 followed by d zeros, then another 1, and then the rest zeros. Now, Z' is greater than Z, and it clearly has the italicized property: For if $N = ab \cdots xy \cdots$, where x is the dth digit, then the calculation of $N \times Z'$ is given by

$$\frac{\begin{array}{r} ab \cdots xy \cdots 0 \cdots 0 \\ + \quad\quad ab \cdots \quad\quad\quad \end{array}}{ab \cdots} \, ,$$

and the dth digit of the product will be x or $x + 1$ depending on whether $a + y$ is or is not greater than 9. The argument for case (b) is similar, except that it involves borrowing rather than carrying.

Notice that this proof is constructive and gives an algorithm for finding the n from the N, but the bound given by the proof is astronomical. For a d-digit number it is roughly 2^{10^d}. In fact, in theory and practice if we look for the first n that works, it turns out that n is on average roughly of the same size as N. For those who are up on such things, we present a *Mathemtica* program, written by Stephan Heilmayr, which finds this smallest n:

$$\text{fract}[x\_]:=\text{If}[x == \text{Floor}[x],1,x - \text{Floor}[x]];$$
$$\text{power2}[x\_]:=\text{Block}[\{Z = \text{fract}[\text{Log}[10,x]],n = 1,$$
$$u = \text{fract}[\text{Log}[10,x + 1]],\text{pow} = \text{Log}[10,2],t = \text{pow}\},$$
$$\text{While}[N[t < Z] \| N[t > = u], \text{pow} = \text{pow} + \text{Log}[10,2];$$
$$t = \text{fract}[\text{pow}]; n = n + 1];n]$$

The fact that n is small is to be expected because one is dealing with a uniform distribution phenomenon. More precisely, the multiples of $\log_2 10$ are uniformly distributed mod 1. Using this fact, one can deduce further properties of the powers of 2. For example, the powers of 2 that contain a million consecutive 7s (or any other given sequence) in their decimal expansions have density 1 in the set of all powers of 2. I thank Jack Feldman for pointing this out to me. What about the stronger statement that almost all, that is, all but a finite number, of powers of 2 contain a million consecutive 7s, or even at least one 7? That is, are there infinitely many powers of 2 that have no sevens at all in their decimal expansions? If you think it unlikely, consider the fact that "practically all" integers contain a million consecutive 7s, meaning that the set of integers with this property has density 1. For note that among all million-digit numbers, the fraction that do not consist entirely of 7s is $(1 - 10^{-6})$, and among all n-million-digit numbers the fraction such that neither their first million, nor second million, ..., $(n - 1)$st million digits consist entirely of 7s is $(1 - 10^{-6})^n$—so you see? On the other hand, of course, there are infinitely many integers with no 7s at all. In other words, statistical information is not really relevant for questions of this sort. As another illustration, recall that in the number 2^{4077} each digit occurs with density approximately one-tenth and presumably this behavior is typical, from which one might be tempted to conclude that perhaps all powers of 2 from some point on contain at least one 7. But again, if we think of the sequence of all integers in their natural order, then by the law of large numbers, those in which the percentage of 7s is less than, say, 9.99 have density zero. Nevertheless, again there are plenty of sevenless integers.

Perhaps the question of whether or not there exist infinitely many sevenless powers of 2 is "unanswerable" in the sense that there is no proof either way from our usual axioms. Would it then, nevertheless, be either true or false? That, of course, depends on one's mathematical philosophy. Or, to fantasize wildly, suppose someone were to show that this question was equivalent to the existence of some unfathomable cardinal? At that point the problem would become theological rather than mathematical—which is probably a good place to stop.

A NONMATHEMATICAL PROBLEM

1. Name the only midwestern state of the United States that does not have an Indian name (midwestern means between the Appalachian and the Rocky Mountains and excludes the southern states on the Gulf of Mexico).
2. Why was this problem included in a column on mathematics?

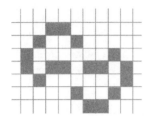

chapter 8

Six Variations on the Variational Method

IDEAS

As a graduate student in physics at the University of Michigan many years ago I had the good fortune to take a course in function theory from Norman Steenrod that pretty much changed the course of my life. My experience in that course plus several private conversations convinced me to switch out of physics and into mathematics. I recall one of our sessions particularly, where, in trying to describe what mathematical research was like, Steenrod said that really there were only about a dozen or so ideas in the whole subject, which people just use over and over again, and once you have mastered these you are, so to speak, in business. I wish now I'd had the presence of mind to ask him for his list of the top twelve. In any case, I expect that on anyone's list, one of the ideas would be the so-called variational method.

THE VARIATIONAL METHOD

The variational method is used to give existence proofs. In trying to prove the existence of an object with certain properties, one picks an object that maxi-

mizes or minimizes some function. The resulting object is then shown to have the desired property by showing that if it did not, one could "vary" the object so that the given function would further increase or decrease. Examples are everywhere dense in mathematics. The most familiar is perhaps Rolle's theorem, and the most historically significant may be Riemann's proof of the celebrated mapping theorem, which he proved by minimizing the Dirichlet integral. As we know, there was a serious gap in Riemann's proof in that he did not show that this minimum existed, and it was not until some years later that Hilbert succeeded in supplying the missing argument.

Of course, the method goes back much further. Perhaps the most basic example is the standard proof of the fundamental theorem of arithmetic, which goes back to Euclid. A crucial step is to show that every two integers a and b have a greatest common divisor, that is, a divisor d of a and b that is divisible by all other divisors of a and b.

THE GREATEST COMMON DIVISOR

Because one is looking for the greatest something, one would expect that this would involve solving a maximum problem. Instead, it turns out that the right approach is to solve not a maximum but a minimum problem, namely, that of finding the *smallest* positive integer d such that $d = ma + nb$, where m and n are integers (there is no problem here about the existence of the minimum, which is, in fact, equivalent to one of the Peano axioms). The fact that any divisor of a and b divides d is now immediate, but it remains to show that d itself is a common divisor of a and b, and this is where the "variation" comes in. If, say, d did not divide a, then by the Euclidean algorithm, there is a positive r less than d such that $a = qd + r$, and r would be a smaller integral linear combination of a and b, a contradiction.

A characteristic feature of this proof is that like many variational proofs, it leads immediately to a simple algorithm for finding the greatest common divisor (gcd) of two numbers.

I have gone over this very familiar proof in such detail to point out that it is more than just an application of the variational method. It is an example of a *duality theorem*. We see that the *largest* integer dividing a and b is the same as the *smallest* positive integer that is an integral linear combination of a and b. Such theorems play a central role in, among other subjects, the theory of linear programming. We will return to this later.

THE SYLVESTER PROBLEM

Another feature of the variational method is the fact that it often leads to very short proofs. A striking example of this is the famous problem posed by Sylvester in 1893: to show that if a finite set S of points in the plane has the property that any line through two of them passes through a third, then the points all lie on a line. Neither Sylvester nor any of his contemporaries was able to find a proof, and it was almost 50 years before the first, rather complicated, proof was published by Gallai. The short proof that follows is now well known. It was discovered in 1948 by L. M. Kelly (*Amer. Math. Monthly* 55, p. 28). Suppose points with the Sylvester property are not collinear. Among pairs (p, L) consisting of a line L through two of the points and a point p not on that line, choose one such that the distance d from p to L is a *minimum*. Let q be the foot of the perpendicular from

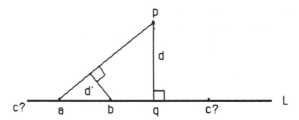

FIGURE 8.1. Kelly's proof.

p to L. Then (variation) by assumption there are at least three points a, b, and c on L. Hence two of these, say a and b, are on the same side of q in the order a, b, q (and c can be on either side) as in Figure 8.1. But then the distance from b to the line ap is d' which is less than d, a contradiction.

Kelly's proof is indeed short—but, here is H. S. M. Coxeter (*Introduction to Geometry*, Wiley, 1961, p. 181): "This matter [Sylvester's problem] of collinearity clearly belongs to ordered geometry. [Indeed, the result is false over the complex numbers or finite fields! You can find an easy 9-point counterexample on the torus.] Kelly's Euclidean proof involves the extraneous concept of distance: it is like using a sledge hammer to crack an almond. The really appropriate nutcracker is provided by the following argument."

Coxeter's lovely proof (which, happily, is also variational) depends on Pasch's Axiom, which in its simplest formulation asserts that it is impossible for a straight line to meet only one side of a triangle. (One way of thinking of this is as a very primitive special case of the Jordan curve theorem. If the line enters the triangle through one side, it must cross another side to get back out again.) The picture for the proof is very much like Kelly's, but this time we pick any point p and find a ray R from it that contains no other points of S but crosses at least one line connecting points of S. Each such line intersects R in some point, so we choose a line L whose intersection q with R is *closest* to p (not in the sense of distance, of course, but as an ordered set on R. That is, there are no other intersection points between P and q.) Now (variation) there must be two points, a and b, of L on the same side of q. We show that there can be no third point of S on the line ap. There are two cases.

Case I. The third point y is between a and p. Then (Fig. 8.2) no matter where c is, from Pasch's Axiom applied to triangle apq, the line cy will intersect R in a point closer to p than q.

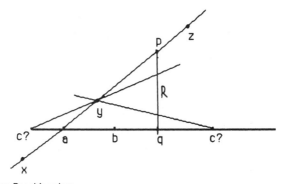

FIGURE 8.2. Applying Pasch's axiom.

Case II. The third point x or z is not between a and p. Then, as before, either bx or bz will intersect R in a point closer to p than q. (See Fig. 8.3.) Voilà!

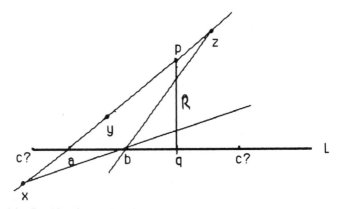

FIGURE 8.3. Applying Pasch's axiom to the other case.

BUT

There are $2n$ points in the plane in general position; n of them are red and n are blue. Show that they span n *disjoint* segments having one end red and the other blue.

Or course, by now you have gotten the message and realize that you should choose segments in a way that minimizes the sum of their lengths. The segments will then be disjoint, for (variation) if they crossed, apply Proposition XX of Euclid's *Elements* (the one about the shortest distance between a red and a blue point being a straight line) to show you could decrease the total length by uncrossing them. Once again, however, we are using the "sledge hammer" of distance on a theorem that is clearly affine (but not projective) invariant. I doubt, however, that this would have bothered the Greeks.

BIRKHOFF'S BILLIARD BALLS

Let T be a convex billiard table with a smooth (C^1) boundary. Then for any $n > 1$, there is an n-bounce periodic billiard-ball orbit (theorem of G. D. Birkhoff).

Proof. Among all inscribed n-gons (whose edges may, of course, cross), choose one whose perimeter is a maximum. This will be a billiard-ball orbit; i.e., the angle of incidence will equal the angle of reflection at each bounce point; for if not, perturb the bounce point on the boundary in the direction of the edge making the larger angle with the tangent. This will increase the perimeter (a nice calculus exercise in what the textbooks used to refer to as "related rates"). ∎

Note that the result is trivial for even-gons, since the ball just bounces back and forth along a diameter of the set.

In Figure 8.4 are shown a pair of 5-bounce Birkhoff billiard-ball orbits on an elliptic table, courtesy of Ben Lotto (as told to *Mathematica*). Interestingly, not much is known about nonsmooth billiard tables, in particular, about polygons. For example, for acute triangles there is always a 3-bouncer with bounces at the feet of the altitudes, the solution this time of finding the inscribed triangle with *minimum* perimeter. (Can you prove it?) It has been shown that periodic orbits exist for polygons whose angles are rational multiples of π. However, apparently nothing is known about obtuse triangles with irrational angles.

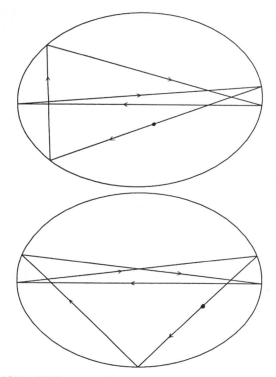

FIGURE 8.4. A pair of Birkhoff billiard-ball orbits on an elliptic table.

THE DESEGREGATION THEOREM

Speaking of short proofs, here is an example that came to me from Donald Newman via Murry Klamkin.

> Given any graph, prove that it is possible to color each vertex either black or white in such a way that at least half of the neighbors of each white (black) vertex are black (white).

I must admit, that I didn't see how to prove this, but Jim Propp did and came up with, not a two-line, but a six-word proof. It is given at the end of the chapter.

THE STABLE-ASSIGNMENT THEOREM

Finally, an example from economics. There are n workers and n employers. If worker i works for employer j, together they will generate a_{ij} units of some good, say bread. Assuming that each employer can hire only one worker, who will work for whom and how will the partners divide up the bread they produce between *wages w* of the worker and *profits p* of the employer? (Wages and profits are paid in bread rather than money to emphasize that we are dealing with intrinsically valuable goods rather than paper and coins.) There is a simple economic "equilibrium" condition that gives the answer. If in an assignment some worker i is earning wage w_i and some employer j for whom i is not working is making profit p_j, then we require that

$$(1) \qquad\qquad w_i + p_j \geq a_{ij},$$

for if the inequality went the other way, then i and j could both get more bread by working together. The question then is whether there will always exist an assignment of workers to firms and a division of the bread from each partnership so that (1) is satisfied. Such an arrangement is called a stable, or *equilibrium assignment.*

Before looking at the existence question, we call attention to a wonderful property of equilibrium assignments that is a special case of what is sometimes called the Fundamental Theorem of Economics. Obviously, for the welfare of the society as a whole, the employer–worker assignments should be such that they produce the maximum possible amount of bread. Such an assignment is called *optimal.*

Theorem. *An equilibrium assignment is optimal.*

Proof. For notational convenience, let us suppose that in the equilibrium assignment worker i works for employer i, so that

$$(2) \qquad\qquad a_{ii} = w_i + p_i.$$

Summing (2) on i gives

$$(3) \qquad\qquad \Sigma a_{ii} = \Sigma w_i + \Sigma p_i.$$

Now consider any other assignment σ where worker i is assigned to employer $\sigma(i)$. From the equilibrium condition (1),

$$(4) \qquad\qquad \Sigma a_{i\sigma(i)} \leq \Sigma w_i + \Sigma p_{\sigma(i)} = \Sigma w_i + \Sigma p_i.$$

The last equation follows because σ is a bijection. From (3) and (4), it now follows that $\Sigma a_{ii} \geq \Sigma a_{i\sigma(i)}$, which is precisely optimality. (It is because of theorems like the above that economists go around extolling the virtues of "market economies.") ∎

We see then that the equilibrium property is related to a very natural maximum problem, which suggests using this maximum problem to prove the existence of equilibrium assignments. April Fool! It turns out as in the case of the gcd that instead one should consider the "dual" minimum problem. Among all $2n$-tuples $(w_1, \ldots, w_n, p_1, \ldots, p_n)$ satisfying the stability condition (1), choose one that *minimizes* $\Sigma w_i + \Sigma p_i$. With these minimizing values of the w's and p's, consider the set S of all pairs (i, j) such that (1) is satisfied as an equation:

$$(5) \qquad\qquad w_i + p_j = a_{ij}.$$

Let us say that if (5) holds for a pair (i, j), then i and j are *compatible*. If it is now possible to choose a subset of n disjoint compatible pairs from this set, then one easily sees that this gives the desired equilibrium assignment.

The punch line in the proof now uses the famous "marriage theorem" of Philip Hall that if no such subset exists, then there must be a "bottleneck," namely, a set of k employers who are compatible with fewer than k workers. But in that case (variation), by the Law of Supply and Demand, one could increase by ϵ the w_i of these "overdemanded" workers, which would decrease the p_j of the k firms, and this would decrease $\Sigma w_i + \Sigma p_j$, a contradiction.

As in our other examples, this proof leads to a good $[O(n^2)]$ algorithm for finding an equilibrium assignment, the so-called Hungarian method, due to Harold Kuhn.

Here is Propp's proof of the Desegregation Theorem:

Maximize the number of interracial neighbors.

Note that once again the proof leads to a good coloring algorithm that terminates in at most E iterations, where E is the number of edges of the graph. Namely, start with any coloring, and if a vertex does not satisfy the conditions, change its color.

(Since this was written, Alexandre Giventhal proposed a three-word proof: Maximize ferromagnetic energy.)

ADDENDUM ON THE VARIATIONAL METHOD

There is, of course, an inexhaustible supply of problems that can be solved by variational methods. Here is a lovely two-examples-in-one case that was suggested to me by Clifford Gardner. It has the special virtue that it provides a simple but elegant application of calculus and would fit in around the third week of a traditional freshman course.

Consider the *discrete heat-flow problem*. Given a graph like the one shown below, where some of the nodes are held at fixed temperatures (3, 5, and -2), the laws of heat flow require that the (steady-state) temperature of every remaining node shall satisfy the *mean value property*, namely, that its temperature shall be the average of the temperatures of the nodes to which it is connected.

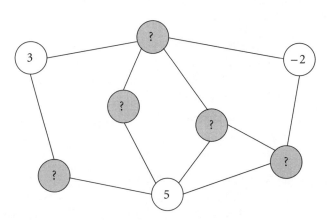

Question. *How do we know that such a set of temperatures will always exist, and if so, are they unique?*

(Of course, this is a problem of solving a system of linear equations, but our students will not get to this until their sophomore year.)

Answer. *Existence:* Let t_i be the temperature at node i, and consider the function $f(t_1, \ldots, t_n) = \Sigma(t_i - t_j)^2$, summed over all pairs of neighboring nodes (this is the *thermal energy* of the system). Choose values of the t's that minimize this function. To see that these values satisfy the desired condition, note that if the temperatures at all nodes except t_k are held fixed at the minimizing values, then t_k must minimize f as a function of one variable, and setting the derivative with respect to t_k equal to zero gives the result.

Ah, you say, but how do we know that the minimum exists? My answer is that mathematics got along for two thousand years without worrying about such questions and there is no reason to inflict them on freshmen. For those who want to be mathematics majors, there will be time enough, when they get to be juniors, to force them, kicking and screaming in some cases, to worry about these matters.

Uniqueness (by the maximum principle!): Suppose there were two sets of temperatures satisfying the mean value property. Then so would their difference, and the difference temperatures at the fixed nodes would be zero. Now consider a node where this difference temperature is a maximum. Then by the mean value property, all of its neighbors must also be at this temperature, and likewise all its neighbors' neighbors, so eventually we will reach one of the fixed-temperature nodes (we assume that the graph is connected); so the maximum is zero. Likewise the minimum.

Q.E.D.

After much cajoling on my part, the retiring director of the Berkeley Mathematical Sciences Research Institute agreed to contribute the following opus which perhaps reflects some of the things he learned during his tenure in office.

The Deep Young Man

A parody of Bunthorne's song in *Patience*, with admiration for, and apologies to Sir William Gilbert and Sir Arthur Sullivan

By Irving Kaplansky

If you wish to cut a path
In the modern world of math
You must spout a lot of things.

You must show you know topology
And étale cohomology,
In physics, mention strings.

You can live a life of pleasant ease
Because a set of PDE's
Is just a space of loops.

You should act a bit convivial
And tell them that it's trivial
If they think of quantum groups.

    Chorus:
    And everyone will beam
    When you tell them of your scheme
    To make them see
    A la Bourbaki
    That if one only tries,
    Then hyperbolic manifolds and affine planes
    Are wavelets in disguise.

And of course you'll give a lecture
On the Poincaé conjecture;
Make it full of double talk.

But don't hint that you can do it,
Others did and lived to rue it,
And use plenty of colored chalk.

You will want to put in lots
Of invariants for knots;
Keep the lemmas coming fast.

Then if their eyes grow bleary,
Just invent a subtle theory
That may settle Fermat's last!

    Chorus (Gilbert's words unchanged):
    And everyone will say
    As you walk your mystic way:
    If this young man expresses himself
    In terms too deep for me,
    Why, what a very singularly deep young man
    This deep young man must be.

chapter 9

Tiling a Torus:
Cutting a Cake

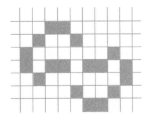

We devote this chapter to some recent results on a pair of fairly well-known problems of recreational mathematics that have been around for quite a while. The first is the problem of tiling surfaces by unequal squares, the second that of devising fair procedures for dividing a cake. The results on tiling are rather definitive, whereas the work on cake-cutting is still in a rather formative stage.

TILING OF SURFACES BY UNEQUAL SQUARES

The question is, or rather was, Which rectangles can be tiled by squares no two of which have the same size? Figure 9.1 is an example of a 32 × 33 rectangle tiled by 9 such squares. This example, apparently discovered by Moron in 1925, appears in Ball's *Mathematical Recreations* and Steinhaus's *Mathematical Snapshots*. In 1940, Tutte, Brooks, Smith, and Stone (*Duke Math J.* 7 (1940), 312–340) were able to show that this is the "smallest" such example, meaning that no rectangle can be tiled in this way by fewer than 9 squares. They also showed, however, that there is exactly one other rectangle, 61 × 69 shown in Figure 9.2, that can also be tiled by 9 squares. The authors were actually seeking and eventually found a

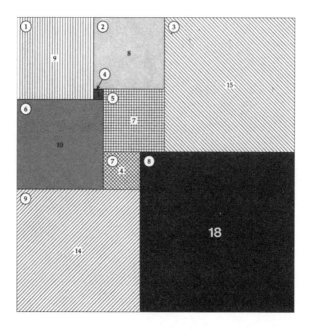

FIGURE 9.1.

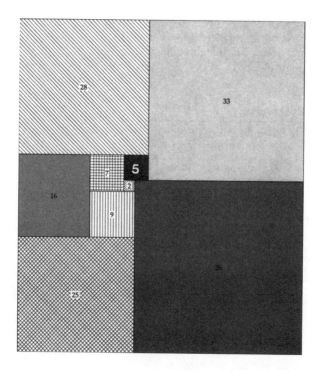

FIGURE 9.2.

square that could be tiled in this way. For an entertaining exposition, see the chapter by Tutte, "Squaring the Square," in Martin Gardner's second *Scientific American Book of Puzzles and Diversions,* Simon and Schuster, 1961.

Now, such a "squaring" of a rectangle can be converted in a trivial way to squaring of the cylinder, torus, Möbius strip, or Klein bottle by the usual identification of opposite sides. But there are also nontrivial squarings of these other surfaces, even *simple squarings,* meaning squarings in which there is no subset of tiles whose union is a rectangle. Until recently, however, it was not known whether there might be squarings of these surfaces requiring fewer than 9 squares. Then, in 1991, Bracewell found a squaring of the Möbius strip using only 8 squares. Very recently the question has been completely settled by S. J. Chapman. Perhaps the simplest but most surprising result is that a 1 × 5 Möbius strip can be tiled by 2 squares, as becomes obvious from Figure 9.3. To accommodate this example, one must extend the notion of tiling to allow the mapping of the squares into the surface to self-intersect on their boundaries.

FIGURE 9.3.

Chapman shows that there are no 3- or 4-squarings of the strip but that there is a unique 5-squaring (Fig. 9.4).

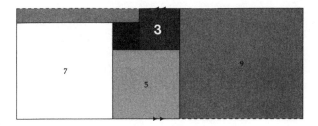

FIGURE 9.4.

For cylinders, the situation is interesting. Again it turns out that 9 squares are necessary. There are exactly two nontrivial 9-squarings of the cylinder, and these use exactly the squares of Figures 9.1 and 9.2 but in a different arrangement. The tiling corresponding to Figure 9.1 is shown in Figure 9.5. Note, for example, that the squares of size 10 and 15 are disjoint in the rectangle, but they are contiguous on the cylinder.

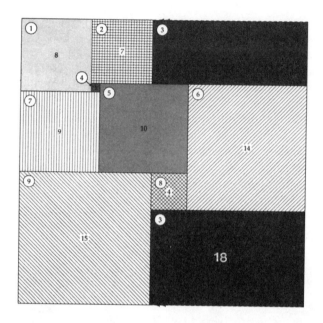

FIGURE 9.5.

The cases considered so far involve surfaces with a boundary, which forces one to orient the squares with one side parallel to the boundary. This is no longer the case with the torus and Klein bottle. If one allows arbitrary orientation, then, in fact, any two squares can tile some torus. Namely, let the squares have sides a and b, and consider the torus obtained from identifying opposite sides of a square of side c, where $c^2 = a^2 + b^2$. Figure 9.6 shows how to do it and at the same time provides a new (?) proof of the Pythagorean Theorem.

A more symmetrical representation is given in Figure 9.7.

If one allows only tiles that are parallel to the sides of the square, then it turns out there are no nontrivial 9-squarings of the torus. Any squaring of the Möbius strip gives a squaring of the Klein bottle. For 6 or fewer tiles these are the only ones, but in the case of 7 or 8 tiles, this is not known. Also it is not known whether there are tilings of the Klein bottle in which the tiles need not be parallel to the sides of the big square.

Chapman's techniques for these problems are quite different from and simpler than those of Tutte *et al.* and depend on a clever encoding of tilings by matrices of 0s, 1s, and −1s.

DIVIDING A CAKE

A cake is to be divided among n of us. We have different tastes. Some of us like the frosting; others are partial to the chocolate filling, etc. Is there a way of giving each of us a piece of the cake such that everyone feels he or she has gotten as desirable a piece as anyone else (such an allocation of pieces is said to be *envy-free*)? Well, it depends. First, the cake must not be too lumpy. If all of us have our hearts set on getting the cherry in the middle, then it is hopeless unless the cherry can somehow be split up among us. This key property, that any piece can be split up into smaller pieces, corresponds to the idea that our tastes are represented by so-called "atomless" measures, countably additive, etc., etc. In this model,

FIGURE 9.6.

FIGURE 9.7.

if by a piece one means any measurable subset, then there is a very strong existence result. Not only is there an envy-free allocation, but there is one in which we all believe that everyone, ourselves included, got exactly one nth of the cake. Otherwise stated, there is a way of cutting the cake into n pieces so that we are all indifferent as to which piece we get. They all look equally delicious. This fact, proved by Dubins and Spanier (1961), is a consequence of a celebrated and moderately high-powered theorem of Lyapunov that states that the range of a vector measure is convex.

Now, as a practical matter, arbitrary measurable pieces of cake may not be so easy to come by. A more down-to-earth model, therefore, has been created by Stromquist (1980), where the cake may be taken to be an interval, and the pieces are required to be subintervals. Perhaps a loaf of bread is a more apt illustration for this case. Using a fixed-point theorem, it is shown that envy-free allocations will always exist. Fixed-point theorems, however, are notoriously nonconstructive, and Stromquist's result gives no indication of how one might arrive at the desired culinary dissection.

A rather different approach to the problem asks not just for the existence of envy-free allocations but for a procedure, or a *protocol* as we shall call it, that leads to such an allocation. The prototypical example is the procedure for the trivial two-person case where one of us divides the cake into two parts and the other chooses the part he prefers. This method has the obviously desirable property that if either of us ends up feeling he has been cheated, he has only himself to blame. It has long been an open problem to try to devise protocols with this property for the n-person case. The best result along these lines is an elegant three-person protocol due to John Selfridge, which will now be described. We denote the players by #1, #2, and #3.

Three-Person Protocol

Step I: #1 "trisects" the cake into 3 parts equally acceptable to him.

$$\#1 \longrightarrow \boxed{A} \quad \boxed{B} \quad \boxed{C}$$

If #2 and #3 prefer different pieces, we are through. Otherwise, say they both prefer A and #2 prefers A to B, which she likes at least as much as C. Thus,

$$\#2 \longrightarrow \boxed{A} > \boxed{B} \geq \boxed{C}$$

Step II: #2 trims a "sliver" (S) from A leaving A′ such that A′ and B are equally acceptable to her.

$$\#2 \longrightarrow \boxed{S}\boxed{A'} = \boxed{B} \geq \boxed{C}$$

Step III: #3 chooses his preferred piece among A′, B, and C.

Case 1. #3 chooses A′. Then #2 chooses B, and #1 gets C (no envy so far).

It remains to divide up the sliver, which is like the original problem, except that now #1 will not be envious even if #3 gets the whole sliver.

Step IV: #2 trisects S.

Step V: #3 chooses, then #1 chooses, then #2.

Case II. #3 does not choose A′. Then #2 gets A′, and the procedure is as before, except that this time #3 trisects and #2 chooses first.

This procedure has a number of nice properties. First, it is economical, requiring at most 5 cuts. Further, like the I-cut-you-choose protocol, it makes minimal assumptions on what the players can do, namely, (1) given any piece and an integer k, it is assumed that a player can divide the piece into k subpieces equally acceptable to him, and (2) if a player prefers one piece to another, she can trim off part of the first piece in such a way that what is left and the second piece are equally acceptable to her. Finally, preferences are required to be only *weakly additive*, meaning that if A, B, X, and Y are disjoint pieces and a player prefers A to X and B to Y, then he also prefers A ∪ B to X ∪ Y.

Up to now, no protocol satisfying the desired conditions has been found, even for the case of four players. However, recent work by Steven Brams and Alan Taylor shows that progress is being made. The authors present what they call a finite algorithm for arriving at an envy-free allocation. Their procedure, however, is quite complicated and seems to require that players be able to measure numerically the value of any piece of cake. Further, even for the four-player case, there is no *a priori* upper bound on the number of cuts that may be required. Thus, for example, if the value of piece A to some player is greater than that of piece B by one part in a million, then it may require a million cuts to arrive at the desired allocation using the proposed algorithm. One might hope that procedures of simplicity comparable to that of the Selfridge protocol could be devised for the general case. On the other hand, it has been conjectured that no such protocol exists, an interesting open problem.

Dividing a Pie: An Unsolved Problem

An allocation may be envy-free but have other undesirable properties. As an example, suppose you and I are to divide a loaf of bread, which again we will take to be an interval, and suppose the loaf is symmetric about its midpoint in both of our measures. Then if we divide it in two at the midpoint, we have a Spanier-Dubins allocation in which we both agree that each of us got exactly half the cake. Suppose, however, that I like crust, so that I particularly want to get the two ends of the loaf, whereas you prefer not to have these parts. Then each of us would be strictly better off if we trisected the loaf in some way and you took the middle part, while I took the two end intervals.

In general, we will say an allocation is *dominated* if there is another allocation that gives all players pieces they strictly prefer. Obviously, it would be desirable for the final allocation to be undominated as well as envy-free. A general question, then, is whether in a given model it is always possible to satisfy both of these conditions. In this connection, recall that in the Stromquist formulation all pieces were required to be subintervals of an interval (unlike the earlier example, in which my allocation was the union of two disjoint intervals). For these Stromquist allocations, we have, in fact, the following theorem:

Theorem. *An envy-free Stromquist allocation is automatically undominated.*

Proof. Let P be an envy-free partition of the interval into n subintervals, and let Q be any other such n-partition. Now, if P and Q are distinct n-partitions of an interval, then there must be some interval I of P that strictly contains some interval J of Q (think about it for a minute). But then whoever gets J in the allocation Q will not be strictly better off than she was under P, for she likes I at least as well as J and she likes the piece she got under P at least as well as I, since P was envy-free. ∎

Which brings us to the problem of the pie. Suppose a pie is to be divided among three people and the pieces are required to be traditional pie portions, namely, *sectors*.

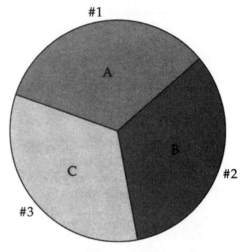

Does there necessarily always exist an allocation that is both envy-free and undominated?

WE ALL MAKE MISTAKES II

In response to my request for examples of mistakes by famous mathematicians several people mentioned a well-known error of Lebesgue. The following nice description was supplied by Doug Lind.

"Lebesgue was trying to prove that the projection of a Borel set in the plane is a Borel subset of the line [Sur les fonctions représentable analytiquement, *Journal de Mathématiques*, Series 6, Volume 1 (1905), page 195]. His argument uses the 'fact' that if you have a decreasing sequence of sets in the plane, then the projection of their intersection is the intersection of their projections (this is contained in the third paragraph starting with "Supposons que e soit F de classe 1 ou 2"). Of course this is wrong (no first-year graduate student should make such a mistake!), and led to the Souslin-Lusin theory of analytic sets." [20-second time-out while the reader finds the obvious counterexample.]

"Lusin even asked Lebesgue to write a preface to his book on analytic sets [*Leçons sur les Ensembles Analytiques et leurs Applications*]. There Lebesgue says that this was the most fruitful error that he had ever committed!"

The Automatic Ant:
Compassless
Constructions

THE INDUSTRIOUS ANT

Most readers of this book are no doubt familiar with John Conway's famous Game of Life. Life is an example of what have come to be called cellular automata. I will discuss here another such automaton, called the ant, and though it is very easy to describe, its behavior is interesting and somewhat mysterious.

The ant lives in the plane, which is divided up into cells by a square grid. There are two kinds of cells, white and black (later we will also introduce gray cells). Initially the ant is sitting on a cell, call it the origin, heading in one of the four compass directions. It proceeds to travel from cell to cell according to the following rule: It moves one cell in the direction it is heading. When it lands on a white (black) cell it rotates its heading 90° to the right (left), and the cell then reverses its color. This is all there is to it. The game is then to start out with some given distribution of black and white cells and see how the ant behaves.

The special case where all cells are initially white and the ant is heading, say, east, is typical of what happens generally. In the early stages of its travels, say the first 500 steps, the ant, at intervals, returns to

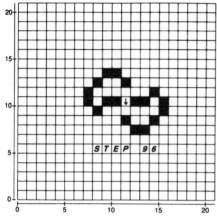

FIGURE 10.1.

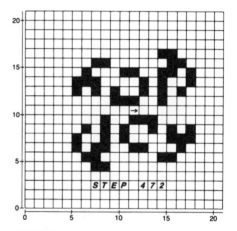

FIGURE 10.2.

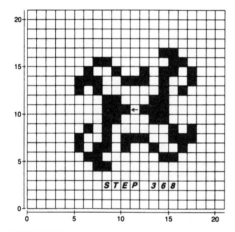

FIGURE 10.3.

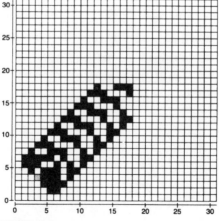

FIGURE 10.4.

FIGURE 10.5.

the origin, leaving behind centrally symmetric patterns of black and white cells as shown in Figures 10.1–10.4. As far as I know, no one has come up with an explanation of why these patterns occur. After a while, however, things become rather chaotic for about 10,000 or so steps, but then the ant suddenly seems to make up its mind where it wants to go and heads off resolutely due southwest, leaving behind the periodic pattern shown in Figure 10.5, which Jim Propp, who first called attention to the phenomenon, calls a highway. On a highway, the ant takes 104 steps, ending up two units southwest of its starting point, and then repeats the process *ad infinitum*.

If the initial position includes some black cells, the ant will, of course, pursue a different course, but in hundreds of experiments it has always ended up building a diagonal highway in one of the four possible directions. Must this always happen? No one knows, but there is one quite charming result, due to L. A. Bunimovch and S. E. Troubetskoy:

Theorem. *An ant's trajectory is always unbounded.*

Proof: First note that the ant rule is reversible. The pattern of black and white cells and the ant's current position and heading determine where it came from. Thus, if a trajectory were bounded, then eventually a black-white pattern would be repeated, and hence, by the preceding observation, the path would have to be periodic, so every cell that was visited would have to be visited infinitely often. Now, the key observation is that the ant's moves are alternately horizontal and vertical. This means that the cells of the plane are partitioned, checkerboard fashion, into h-cells, which are always visited horizontally (from the right or left), and v-cells, which are always visited vertically (from above or below). Now consider a "maximal" cell M that was visited by the ant, meaning a cell such that no cell above or to the right of it has been visited. Suppose M is an h-cell. By maximality, it must have been entered from the left and exited downward, so the cell must have been white. But then it turns black, so on the next visit from the left, the ant must go up, contradicting maximality. If M is a v-cell, the argument is similar.

Neat, don't you think?

A variation on the game is to introduce a third type of cell, a gray cell, with the property that when the ant lands on such a cell it simply continues in its current direction. Gray cells do not change their type but remain gray forever. Note that in this model the proof of the unboundedness theorem fails because the partition into h- and v-cells is no longer valid; and indeed Cohen has found a fairly simple example of an initial configuration, shown in Figure 10.6, yielding a trajectory that repeats every 52 steps.

One gets some rather pretty patterns by starting with an initial line of gray cells, leaving all remaining cells white. Figures 10.7–10.10 show how, initially, the ant behaves like a spider spinning a web, but gradually asymmetries appear, and highway construction begins at about step 9000.

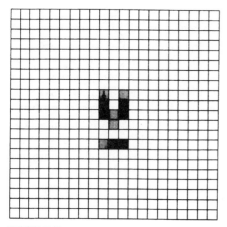

FIGURE 10.6.

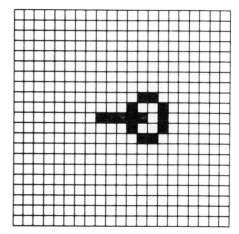

FIGURE 10.7.

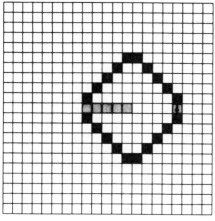

FIGURE 10.8.

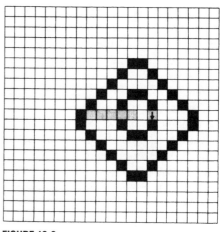

FIGURE 10.9.

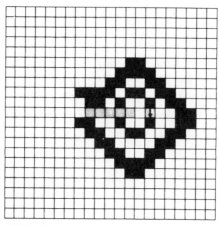

FIGURE 10.10.

The ant was invented by Chris Langton, who studied its behavior in some detail, including cases where several ants are moving simultaneously.

STRAIGHTEDGE CONSTRUCTIONS

The best way to learn a new subject, as everyone knows, is to teach it. Sometimes, if things go well, one even ends up making an original contribution. I was surprised, however, when this happened to me recently in a course I was designing for fourth- and fifth-grade schoolchildren. It was to be about geometric construction, not with the traditional ruler and compass, but with a "markable" straightedge. That is, the only equipment supplied was to be a pencil, an eraser, and a rectangular strip of white cardboard on which one could make (and erase) marks.

By way of a warm-up, the students were to be asked to use this equipment to determine which of the segments in the Figures 10.11–10.13 was longer. As a variation, they could then be asked to make a dot at the point that they considered to be the midpoint of the arrow shaft in Figure 10.14, and then use the straightedge to find out whether their

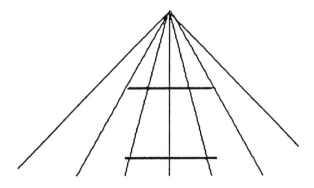

FIGURE 10.11.

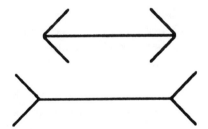

FIGURE 10.12.

FIGURE 10.13.

FIGURE 10.14.

guess was to the left or right of the true midpoint. The idea here is to see whether the students are able to use the straightedge suitably marked to compare the distance of their guessed point from the two ends of the arrow. The well-known optical illusions used here are intended to convince them that there is some point in making these measurements with care.

This last exercise leads to the first example of a construction problem. Given a line segment, how can we use the straightedge to find its true midpoint? My proposed solution, which will probably have to be shown to the students, is to construct perpendiculars of equal length at the two endpoints of the segment and connect them as shown in Figure 10.15. (I am assuming at this point that the straightedge is a true rectangle, so that constructing perpendiculars is immediate. I will return to this later.) This can lead to some significant discussion. One asks the students to explain how they know that this construction really does find the midpoint, which leads into the important subject of symmetry. (As an interesting digression, one may change the rules and allow the students to use an ordinary sheet of paper along with their pencils to find the midpoint and see how many of them come up with the idea of marking and folding the sheet appropriately. In my limited experience a child is just as likely as an adult to figure this out.)

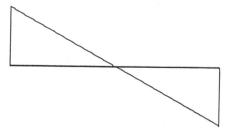

FIGURE 10.15.

The next step is to divide a segment in 3 equal parts using the construction in Figure 10.16. Again there is the opportunity for some pertinent discussion as to why one believes that the length of the right-hand segment is one-third of the total length. After this, the students could be asked to divide a segment into 5 equal parts on their own, and so on.

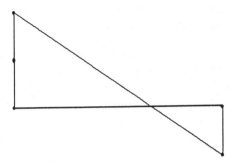

FIGURE 10.16.

One can now do a variety of things. A nice exercise, for example, is to give the students a horizontal line segment on a sheet of paper and ask them to construct the equilateral triangle having this segment as a base. The idea is to construct the perpendicular bisector of the segment and then mark the end points of the segment on the straightedge (Fig. 10.17).

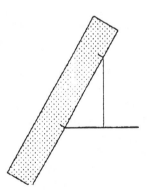

FIGURE 10.17.

It is now necessary to be a bit more explicit about which operations are permitted with the marked straightedge. As illustrated by this last example, one can find the intersection of a given line with a circle of given center and given radius, even though one cannot draw the circle. Otherwise stated, given a line L and a point P not on L and given two marks, A and B, on the straightedge, one may obtain a new point Q on L by placing the straightedge so that A falls on P and B lies on L, assuming, of course, that this is possible. Indeed, all the classical ruler-and-compass constructions are possible because it is not hard to see how one can construct "square roots of segments."

The next exercise was to show how to bisect an angle (by making marks on each ray at equal distances from the vertex and constructing the perpendiculars at these points and connecting their intersection to the vertex). I was about to write some remarks to the effect that although, as was shown earlier, we can "trisect" segments with our straight-edge, the analogous construction for angles has been proved to be impossible, and then I realized that this assertion was false! Herewith is a trisection, attributed to Pappus, which uses only a marked straightedge (Fig. 10.18).

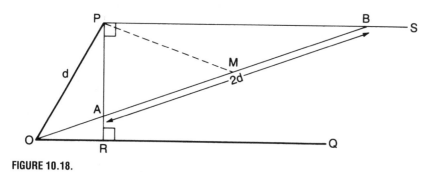

FIGURE 10.18.

The angle BOQ is one-third of POQ.

Proof: As indicated, the length of segment AB is twice that of OP. Now draw the line from P to the midpont M of AB. Then the length of PM is d because—and you can take it from there.

So the question is, what is it that the marked straightedge can do that a ruler and compass cannot? The construction above proceeds by choosing the point P arbitrarily and then dropping the perpendicular from P onto OQ. Now one makes marks A and B, $2d$ units apart on the straightedge, and one must place the straightedge so that (1) it passes through the point O, (2) the mark A lies on PR, and (3) the mark B lies on PS. This is the crucial maneuver. More generally, one is given a line L and a point P not on L and two marks X and Y on the straightedge. Now consider the locus traced out by Y when the straightedge is placed so that it passes through P and the point X lies on L. This locus, known as the conchoid of Nichomedes, is given by a 4th-degree equation. Typical graphs are shown in Figure 10.19. Of course, one is not allowed to draw the conchoid, but one can find its intersection with a line, as in the Pappus construction. Manually, this amounts to placing the straightedge so that two given marks X and Y lie on a pair of given lines and then "sliding" the straightedge, keeping the marks on the lines, until it passes through a given point, a fairly easy operation to perform.

Now the interesting fact is that this conchoid maneuver allows one not only to trisect angles but, in fact, to find the roots, real or complex, of any polynomial of degree at most 4. The trick is to show that it is possible to construct cube roots of positive numbers. Figure 10.20 shows how to do it. The claim is that **x** is the cube root of **a** (boldfaced letters are used here for lengths). This can, of course, be proved analytically, although the calcu-

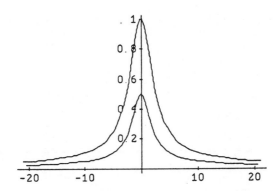

FIGURE 10.19.

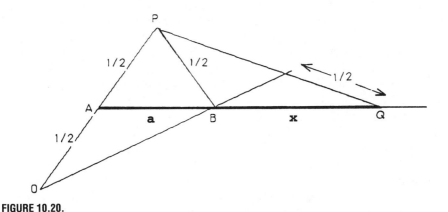

FIGURE 10.20.

lations get rather tangled if one does not set things up conveniently. There is a slick proof using the theorem of Menelaus.

In making the construction, one chooses the unit of length, in this case *OP.* Now lay off the segment *AB* of length **a,** construct the isosceles triangle *APB,* the point *O,* and the prolongations of the segments *AB* and *OB.* With marks a distance ¹/₂ apart on the straight-edge, one then uses these two lines together with the point *P* to locate the point *Q* by means of the "conchoid maneuver."

Finally, using trisections and cube roots, one can take cube roots of complex numbers (de Moivre's theorem), and these together with square roots are sufficient to find the roots of any polynomial of degree at most 4.

The point of all this is the observation mentioned earlier that using this seemingly primitive tool one can make constructions that are impossible with the classical ruler and compass. Thus, for example, from Galois theory, one sees that it is possible to construct a regular heptagon as well as a 13-gon and a 19-gon, but not an 11-gon. In general one can construct a *p*-gon for those primes *p* such that $(p - 1)$ has only 2 and 3 as prime factors.

Returning now to the subject of construction perpendiculars, one may ask whether the straightedge needs to be equipped with built-in right angles. Suppose, for example, it looked like this:

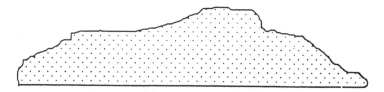

Using only such a straightedge, can one erect perpendiculars? It turns out that one can. To erect a perpendicular to a given line at a given point ... but on second thought, let me leave this construction as an exercise. My solution involves drawing two auxiliary lines and two auxiliary points, and making two marks on the straightedge. On the other hand, to drop a perpendicular from a given point to a given line costs me four lines and three straightedge marks, but no additional points. As a further diversion, suppose one is to draw a line through two points A and B whose distance apart is greater than the length of the straightedge. I will assume that by taking careful aim one can draw a line from A that misses B by no more than the length of the straightedge. I then have a somewhat cumbersome construction.

Needless to say, none of the foregoing results is new, with the possible exception of the very elementary facts of the preceding paragraph. One source for the material is an interesting book by the Danish geometer J. Hjelmslev called *Geometriske Eksperimenter,* which has been translated into German as a *Beiheft* to the *Zeitschrift für Mathematischen und Naturwissenschaftlichen Unterricht,* 1915. According to Hjelmslev, the cube-root construction given above appears without proof in Newton's *Arithmetica Universalis* (Cambridge, 1707), but Hjelmslev states that Newton's proof is only a slight variation on one given by Nichomedes.

I suppose one could carry this further and consider constructions that can be made with both a marked straightedge and a compass. This would produce loci of degree at most 8. Perhaps such studies have been made.

chapter 11

Games:
Real, Complex,
Imaginary

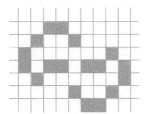

GAMES PEOPLE PLAY

The subject of game theory is by now a fairly substantial branch of mathematics, treated in dozens of books, hundreds of research papers, and at least two full-time journals. Game theory comes in many quite different flavors, including economic, political, military, and, of course, recreational; but all of these branches have one thing in common. In all but very rare cases, the games analyzed are not traditional existing games but rather new games, invented not to be played but to be analyzed. There are a few exceptions like the game of Dots and Boxes, which predates game theory. Also very recently Elwyn Berlekamp has been applying some of the most sophisticated ideas of combinatorial game theory to certain positions in the game of go. However, this work has little to do with actually playing go, in the same way that chess problems have little to do with playing chess.

All of this is by way of leading up to the fact that in contrast to the remarks above, Michael Paterson and Uri Zwick have recently succeeded in giving an almost complete analysis of the familiar game of *Concentration*, which many readers have no doubt played as children or, perhaps later, with children. A deck consisting of *n* pairs of matched cards is shuffled, and the cards are spread out face down on a table. A player turns over a card and then a second card. If the two cards match, the player keeps the pair and gets another turn. If not, the chosen cards are again turned face down, and the turn passes to the opponent. The winner is the player who accumulates the largest number of pairs. Because both players see all cards that have been turned over, success at the game depends on skill in remembering the location of the cards that were previously turned. Thus, people with photographic memories should do well. But what if both players have perfect memories? Would it then be just a matter of the luck of the draw? The author's first striking observation is that in fact, strategic behavior is possible, and sometimes it is advantageous to make a deliberately "stupid move." We illustrate.

Consider a deck containing four pairs, say aces, kings, queens, and jacks. On his first move, your opponent turns over an ace and a king. You then turn over the other ace, collect the pair, and play again. This time you turn over a queen on your first draw, so the state of the game is represented below.

The dotted card indicates that the king is face down but its position is known to both of you. In fact, since the players have perfect memory, we may as well assume that cards that are turned over remain face up. The queen is face up, and the position of the remaining cards is not known. Now you must turn over a second card, so naturally, you should turn over one of the unknown face-down cards. Wrong! The proper move is to turn over the king, even though you know that this will gain you nothing and will pass the turn to your opponent. Why is this? All right, if you turn up another card, then with probability $1/2$ it will be a jack in which case your opponent will get all three pairs. So your expected loss is $3/2$. On the other hand, if you do not draw the jack, you are equally likely to draw the king or queen, and from symmetry one sees that your expected loss in the first case is the same as your expected gain in the second. So your overall expected loss is $3/2$. On the other hand, suppose you turn over the card you know to be the king. Now with probability $1/2$ your opponent draws the jack and will win or lose the three pairs depending on whether or not he then draws the other jack. So his expectation is $-1/2$. If he draws the king or queen, he takes the pair and is then twice as likely to win as to lose the other two pairs, giving him an expectation of $5/6$. So his total expected gain, which is your expected loss, is $1/3$. To summarize, if you are playing for a dollar a pair, then the "stupid move" costs you only 33 cents as opposed to the dollar and a half you would expect to lose if you played in the usual way (from now on in all of the analyses, it is assumed that you are interested in maximizing your expected winnings. This is the difference between the number of your pairs and your opponent's pairs).

Of course, the common sense behind the above calculations is fairly clear. In deciding whether or not to turn over a second card, you are weighing the chance of getting a match, on the one hand, against the information you give your opponent in case you fail, on the other. As the simplest example of this, consider a deck consisting of only two pairs. Then it is clearly a disadvantage to play first, because you have only one chance in three of getting a match, and if you do not, you lose both pairs. Is it then a disadvantage to go first for a deck of any size? The answer is yes for decks of 3, 4, and 5 pairs, but for 6 and 7 pairs the first player has the edge. From then on, the advantage is for the first or second player according to whether the number of pairs in the deck is odd or even—a fact that seems far from obvious.

Returning to the question of strategic play, the best strategy depends on the assumption one makes about one's opponent. In the example above, we considered a "naive" opponent, meaning one who always turns over a second card when her card doesn't match a face up card. We will refer to this as a *2-move*. In this case, it turns out that one should almost always play the stupid move, which we will refer to from now on as a *1-move*. However in some situations one should play a "superstupid move." Thus, in the four-pair game, if your opponent starts by turning over the ace and king, the best thing you can do is turn them up a second time. This of course amounts to passing, and, as was seen in the previous paragraph, this may well be the best thing to do. We will refer to this as a *0-move*. Further, because we are assuming perfect memory, we may as well assume that cards that have been turned over remain face up.

```
n=  1 28
n=  2 028
n=  3 1028
n=  4 2100
n=  5 02100
n=  6 102110
n=  7 2100110
n=  8 02110110
n=  9 210110110
n= 10 0210110110
n= 11 10210111110
n= 12 210111111110
n= 13 0210111111110
n= 14 10211111111110
n= 15 211011111111110
n= 16 0211111111111110
n= 17 21011111111111110
n= 18 0210111111111111110
n= 19 1011111111111111110
n= 20 21011111111111111110
n= 21 021111111111111111110
n= 22 1011111111111111111210
n= 23 211111111111111111111210
n= 24 02111111111111111111111210
n= 25 2111111111111111111111210
n= 26 2111111111111111111111210
n= 27 1011111111111111111111111210
n= 28 21111111111111111111111111210
n= 29 0211111111111111111111111210
n= 30 2111111111111111111111111111210
n= 31 21111111111111111111111111111210
n= 32 211111111111111111111111111111210
n= 33 21111111111111111111111111111111210
n= 34 2111111111111111111111111111111111210
n= 35 21111111111111111111111111111111111210
n= 36 211111111111111111111111111111111111210
n= 37 2111111111111111111111111111111111111210
n= 38 21111111111111111111111111111111111111010
n= 39 211111111111111111111111111111111111111010
n= 40 2111111111111111111111111111111111111111010
```

The *k*th entry in row *n* tells which move to play in a game with *n* pairs and (*k* − 1) face-up cards.

The general optimal strategy against a naive player turns out to be the following: Aside from sporadic exceptions for decks of fewer than 30 pairs, one should always play a 1-move when the number of pairs has the same parity as the number of face-up cards. When these parities are different, one still plays a 1-move until the proportion of face-up cards is very large. At some critical point, one then in some cases plays an isolated 2-move followed by 0-moves. The picture is given by the table above, where it is assumed that 0- and 1-moves are permitted even at the start of the game, when no cards have been turned over.

The work of Paterson and Zwick is concerned not with the naive opponent but rather with two equally sophisticated players. In this case the optimal strategy is even easier to describe. With the exception of the six-pair deck, the rule is that if the number of pairs and face-up cards has the same parity, one plays a 1-move. For the other case, one plays a 2- or 0-move depending on whether the number of face-up cards is more or less than two-thirds of the number of pairs. Of course, a 0-move by either player means that the game will terminate, because otherwise, both players would continue playing 0-moves forever.

The story of the Paterson-Zwick result is intriguing. First, it was a happy circumstance that the pattern of the optimal strategy turned out to be so easy to describe. Of course, this pattern could not have been discovered without the aid of high-speed computation (surely no one could possibly have conjectured such a pattern *a priori*). On the other hand, given the availability of such computation, it was a relatively routine matter to obtain as much data as one needed. Thus, the optimal move for a given configuration of face-up and face-down cards depends on the optimal move for configurations with fewer face-down cards and it is not hard to formulate the appropriate recursions in this way. What is rather remarkable, however, is that the authors were able to prove with complete rigor that the strategy described above is indeed optimal. Further, the proof is extremely involved, running some fourteen pages broken up into seven sections. Perhaps most noteworthy of all is the fact that the proof makes heavy use of computer-aided symbolic computation. In the authors' words, "It is doubtful whether this analysis could have been carried out without resort to experimentation and a substantial use of automated symbolic computations."

The observations above should not be taken to mean that the whole project was handled by computers and that all the investigators had to do was feed in the data. On the contrary, the main challenge in attacking this problem was to put it in a form that made the decisive symbolic computations possible. Some of the section headings, "Operator Notation," "Bootstrapping," "Boundary Layer Influence," indicate some of the concepts involved in this effort. Here then is one more example to suggest that mathematicians may want to rethink their notions of what it means to "do mathematics" in the age of high-speed computation.

Meanwhile, despite these new discoveries, children and parents will no doubt continue to play *Concentration*, because for better or worse, there seems to be no indication that humans are about to develop perfect memory.

GAMES PEOPLE DO NOT PLAY

Here is a memory game of a different sort. I have an infinite deck of cards, and on my first move I hand you a finite subset of them. You are then allowed to discard one card, after which I hand you another finite subset of the remaining cards, and you again make a single discard. The game continues in this way for a countable number of moves, and if at the end your hand is empty, you win, and otherwise I do. Simple.

Now, as described above, the game is not very interesting, because in fact, you have an easy winning strategy. You simply keep track of the cards as I hand them to you and proceed to discard, first the cards I gave you on my first move (in any order), then those I gave you on my second move, and so on. Thus, by the end of the game you will have gotten rid of them all. Notice, however, that to execute this strategy you must have a very long memory. For example, if I give you a million cards on each move, you will continue to fall further and further behind, and by the time you have gotten rid of the first hand I gave you, you will be holding 10^{12} cards. In other words, to play the simple strategy, you must have an unbounded memory. Let us, therefore, go to the other extreme and suppose that you have no memory at all, but on each turn know only the names or numbers of the cards you are currently holding (note that this sort of information is good enough for finite games, like chess, checkers, or tic-tac-toe, where to decide on your next move, the only thing that matters is the current position of the board and not how the position was reached). If the deck is countable, you are still in good shape. Before the game starts, you enumerate the deck in some fashion, and then on each turn you simply discard the card in your hand with the smallest number. What happens, however, if the deck is uncountable?

Theorem 1. *For an uncountable deck in the zero-memory game, there is no winning strategy for the second player.*

Proof: A strategy in the memoryless game is simply a function that assigns a card (your discard) to each finite set of cards (your hand). In particular, therefore, it assigns to each pair of cards the one that is to be discarded. Let us then define for each card x the set D_x consisting of all cards $y \neq x$ such that if you hold $\{x, y\}$ you will discard y. Now, to show that you have no winning strategy, it will suffice to prove that there is some x' such that $D_{x'}$ is infinite, because then it will contain some infinite sequence $y_1, y_2, \dots$ So I will give you $\{x', y_1\}$ on my first move and $\{y_n\}$ on my nth, and you will never get rid of x'. To complete the proof, we construct such an x'. Namely, choose a countable set C of cards x, and suppose all D_x are finite for x in C (if not we are through). Then clearly the union U of C with all the D_x for x in C is countable. But then any x' in the complement of U (nonempty because the deck is uncountable) has the property that C is contained in $D_{x'}$, and this is the desired construction. ∎

So you have no winning strategy in the zero-memory game. However, it turns out that to win you do not need unbounded memory. All you need to remember is your last discard (which is perhaps left face up on the top of the discard pile) and once again you have a win if (and only if!) the axiom of choice is permitted.

Theorem 2. *Using the axiom of choice, there is a win for the second player provided that she knows the cards in her hand and also her last discard.*

Proof: The winning strategy is simple. Well-order the deck. Then on your first move, discard your highest card. Thereafter, (a) discard the highest card that is lower than your last discard if there is one. Else, (b) discard your highest card. To see that this works, note that as long as you are in case (a), the last discard will be strictly decreasing as a function of time, and hence by the well-ordering property the sequence must be finite. So case (b) must eventually occur. Now look at some card x in your hand. If it is less than the current discard, then by the observation above, you will surely discard it at some future time. If it is greater than the current discard, then when case (b) occurs, you will discard x or some card greater than x, and then the previous argument again applies. ∎

A third result, which also uses well-ordering, shows that the second player also has a winning strategy if at each move she is allowed to see the cards in her hand and also the set of all the cards that have been discarded.

Theorem 1 is a special case of a general result of Martin Furer and Ernst Specker, whereas Theorem 2 and the third result mentioned are due to Richard Laver and Krzystztof Ciesielski. As far as I know, none of this work has been published.

GAMES PEOPLE *COULD* PLAY

The following game, christened Chomp by Martin Gardner, is a variation on Nim.[*] It has been around for quite a while, but some recent results lead to a number of intriguing questions. The game is easily described. One begins by laying out a rectangular array of "cookies." Players move alternately by choosing a cookie and then removing all cookies above and to the right of it. Figure 11.1 shows an initial 3×5 Chomp position and the position after a possible first move in row 3, column 4.

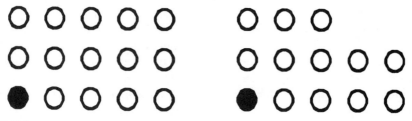

FIGURE 11.1.

The loser is the player who is forced to pick up the poison cookie in the lower left-hand corner. The essential feature of the game is that with optimal play it should always be won by the first player. The argument is illustrated by Figure 11.2.

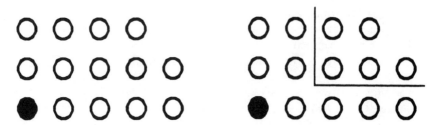

FIGURE 11.2.

---

[*]See Appendix 1.

If removing the upper right-hand cookie (Fig. 11.2a) gives player 1 a winning position, then the assertion is proved. If not, then it gives a losing position, and player 2 has a response that gives her a winning position (Fig. 11.2b); but any such move by player 2 is a move player 1 could have made himself on the first move.

It is to be noted that this argument is completely nonconstructive, asserting only that a winning strategy for the first player exists but giving no indication of how to find it. Indeed, except for two special cases, almost nothing is known about the structure of these winning strategies. The exceptions are the $2 \times n$ (and $n \times 2$) games, where one easily sees that the first player can always give his opponent a position in which the bottom row has one more element than the top row (Fig. 11.3):

FIGURE 11.3.

Hence he eventually wins. The other example is the $n \times n$ case, where the first player plays as shown (Fig. 11.4) and thereafter plays the same move as his opponent but reflected in the diagonal.

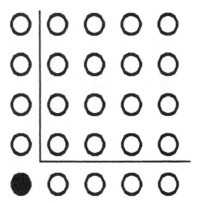

FIGURE 11.4.

As an illustration of our ignorance, experimentally it seems that in $3 \times n$ Chomp the winning first move is unique. The winning move for 3×5 is shown on the right in Figure 11.1, but these winning first moves have no apparent pattern. It was originally conjectured that the winning first move was always unique, but the 6×13 game turned out to be a counterexample in which there are two winning first moves.

Chomp can also be played with an infinite number of cookies. The simplest example is the $2 \times \omega$ case (Fig. 11.5). Note that even though there is an infinite number of cookies, the game can last only a finite number of moves. In fact, this is the first example of a game that is a win for the second player. It is easy to see that no matter what the first player does, the second player can move to give a position like that of Figure 11.3. It then follows that the $\omega \times \omega$ game, shown in Figure 11.6, is a first-player win. By choosing the element in row 3, column 1, he can give the opponent the $2 \times \omega$ game.

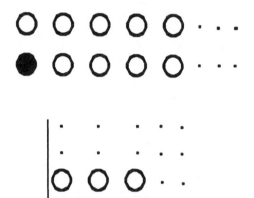

FIGURE 11.5.

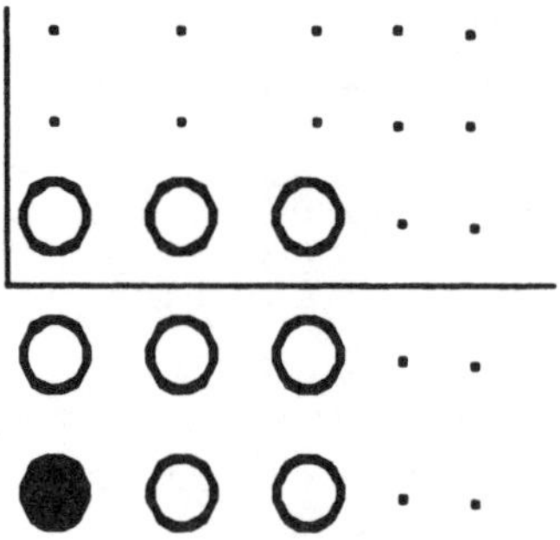

FIGURE 11.6.

In an obvious way, the $\omega \times \omega$ game can be represented by the set of all lattice points in the positive quadrant of the plane, and this can be generalized to three or more dimensions. The game on the positive octant in 3-space remains unsolved; is it a first- or second-player win? The new results alluded to earlier reduce this question to a problem in the arithmetic of transfinite ordinals, as will now be explained.

First, we note that Chomp can be played with arbitrary ordinals; thus, given any ordinals α and β, we may consider $\alpha \times \beta$ Chomp. Because of the well-ordered property of ordinals, it follows that any play of these games must terminate in a finite number of moves. Similarly, one could also consider games played on the product of three or more ordinals.

Next, a game position that is losing for the player who is about to move has been called a *P-position* by Berlekamp, Conway, and Guy (*Winning Ways*, Academic Press, 1982), where the *P* indicates that the game is a win for the previous mover. Figures 11.3, 11.4, and 11.5 are *P*-positions for Chomp. The following fact, communicated to me by Scott Huddleston, is a special case of what he suggests might be called the Fundamental Theorem of Chomp. We state it as follows:

Theorem. *For any integer n, there exists a unique ordinal α such that the $n \times \alpha$ rectangle is a P-position.*

Proof: We first note that the uniqueness of the critical ordinal is immediate. If α gives a *P*-position with $\alpha < \beta$, then $n \times \beta$ is not a *P*-position because the first player can move to change the position to $n \times \alpha$.

We have already seen that if $n = 2$, then $\alpha = \omega$. Huddleston is convinced that the rectangles $3 \times \omega^\omega$ and $4 \times \omega^2$ are P-positions, but I have not seen proofs of these intriguing assertions, nor has he supplied a proof of the Fundamental Theorem. But George Bergman has, and the following is a mild modification of his argument. We will show, in fact, that the critical ordinal α must be a countable ordinal. So from now on all ordinals considered are assumed countable.

Instead of cookies, it is better to think of a Chomp position as a nonincreasing sequence of ordinals $(\alpha_1, \ldots, \alpha_n)$. For any integer $0 < m < n$ and ordinal α, define $f(\alpha, m)$ as the unique ordinal β, if it exists, such that $(\beta, \beta, \ldots, \beta, \alpha, \alpha, \ldots, \alpha)$ is a P-position, where β occurs m times (this is a position after a first move in an $n \times \beta$ game in which β has been replaced by α in the top $(n - m)$ rows). Let $g(\alpha) = \sup f(\xi, m)$ for all $\xi < \alpha$ and $0 < m < n$. Because the set of these (ξ, m) is countable, $g(\alpha)$ is a countable ordinal. Now define the sequence of ordinals (α_i), where $\alpha_1 = 1$ and $\alpha_i + 1 = g(\alpha_i)$, and let γ be the smallest ordinal greater than all α_i. If $n \times \gamma$ is a P-position, the theorem is proved. If not, there is a winning first move, which gives a P-position $(\gamma, \ldots, \gamma, \xi, \xi, \ldots, \xi)$ with $\xi < \gamma$, where γ occurs m times. Let us see whether $m > 0$ is possible. Then γ would be $f(\xi, m)$, and this would contradict the definition of γ. Therefore, the winning move must give a position $(\xi, \xi, \ldots, \xi)$, so the $n \times \xi$ game is a P-position. ∎

Note that like Chomp proofs in general, this one is completely nonconstructive, giving no clue for finding the critical ordinal.

A slightly more complicated argument shows that for any pair of ordinals α and β there exists a unique ordinal γ such that the $\alpha \times \beta \times \gamma$ game is a second-player win (thus, a P-position). Huddleston has asserted that $2 \times 2 \times \omega^3$ is a P-position. The problem would be solved if one knew for which ordinal α the $\omega \times \omega \times \alpha$ game is a P-position. A possibly simpler question is to decide whether the $3 \times 3 \times \omega$ game is a first-player win. Any takers?

AN OLD STORY IN A CONTEMPORARY SETTING

A traditional women's college is considering admitting men. The chairman of the Board of Trustees thinks it would be a good idea, but the college president disagrees. "If you admit men," she says, "more than half of the women students will leave before the end of the year." As a compromise, they agree to admit only 1% men on an experimental basis.

The end of the year rolls around, and at the Trustees' meeting the chairman announces triumphantly that the experiment has been a success. True, he admits, there were some defections by women, but the drop in percentage of women students was only 1%, from 99% at the start of the year to 98% at the end.

"What did I tell you!" says the college president.

chapter 12

Coin Weighing:
Square Squaring

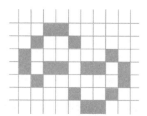

THOUGHT LESS MATHEMATICS

Donald J. Newman

One type of problem that we all "teethed on" in our mathematical youth was the so-called weighing problem. We learned therein the valuable lesson of "branching" procedures: if this and this happens, then we do that and that, but if it does not happen, then instead we do such and such.

These "branching" procedures emerged as a fundamental method in weighing problems, perhaps the right method for problems in general. Our minds were even tempted to go further. This might be the right path to follow for mathematics in general. And if it is right for mathematics, then it might be the right way to think altogether! WOW.

In an article in the *New Yorker,* Jeremy Bernstein pointed with admiration to the use of this branching reasoning as an index of real mathematical talent. The example he chose was the famous 12-coin problem, and the solver he pointed to was the then Harvard undergraduate Charles (Ariel) Zemach.

So, this branching reasoning appeared to be fundamental and nearly universal.

Our purpose, however, is to deflate this notion! We illustrate with the 12-coin problem itself and a few other examples and hope to convince the reader that this branching reasoning is perhaps never needed. Any time there is a solution using branching, there is another one that does not use it (and is, as a result, cleaner and simpler).

The 12-Coin Problem

Twelve identical-looking coins are given, and we are told that one of them has a weight different from the other 11. The problem is to determine which coin it is and whether it is heavier or lighter, in only three weighings of these coins on a balance scale.

Note first that even to describe a solution after one has found it seems to require branching. The *outcome* of a weighing is that the left tray of the balance goes down or up or remains level. We encode these outcomes by 1, -1, and 0 respectively. Each weighing involves picking a pair of subsets L and R of the same cardinality and putting them on the left and right trays, respectively. The "flow diagram" for a weighing procedure looks like this:

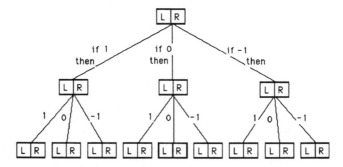

where for each L and R one would have to list the elements of the appropriate subsets. Compare this to the following nonbranching instructions. Let the coins be labeled A, B, ..., L. Then do

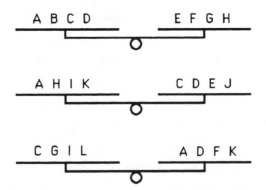

One easily verifies that this works. Namely, if A is the phony coin and it is heavy, then the outcome will be 1, 1, -1, and if it is light, the outcome will be -1, -1, 1. If B is the culprit and is heavy, then the outcome will be 1, 0, 0, and if it is light—but wait a minute. We do not have to go through all this. All we have to do is notice that no two coins have the same or opposite "itineraries"; that is, no two coins are always on the same tray or always on opposite trays. Therefore, for each of the 24 possible *states* (that is, which coin is counterfeit and whether it is heavy or light), there will be a different outcome. So given the outcome, we will know the state. For example, if we know that the outcome is −1, 0, 1, then the coin *G* must be heavy. Note, by the way, that it makes no difference in which order the weighings are performed. Also note that we can solve a slightly harder problem in which we allow the possibility that none of the coins is counterfeit, which will be true if and only if the outcome is 0, 0, 0.

But the question now is what sort of ingenuity was required to find these three weighings. The answer—*none.* We let the solution give itself. The real message we wish to impart then is that old, complicated, clever solution was a waste of effort, an incorrect attitude! The new solution reasons backward from the 27 outcomes to the 24 counterfeit possibilities. Here's how it goes.

First, we make a list of 12 different outcome vectors, such that no outcome and its negative are on the list. A simple way to do this is to list the lexicographically positive vectors in lexicographic order, as in the columns of the following table.

A	B	C	D	E	F	G	H	I	J	K	L
0	0	0	0	1	1	1	1	1	1	1	1
0	1	1	1	−1	−1	−1	0	0	0	1	1
1	−1	0	1	−1	0	1	−1	0	1	−1	0

Now, for the procedure to work, we must have the same number of 1s and −1s in each row. The bottom row is correct as it stands. Reversing the sign of column *C* fixes the middle row, and reversing columns *F*, *H*, *J*, and *L* takes care of the top row. So we have

A	B	C	D	E	F	G	H	I	J	K	L
0	0	0	0	1	−1	1	−1	1	−1	1	1
0	1	−1	1	−1	1	−1	0	0	0	1	−1
1	−1	0	1	−1	0	1	1	0	−1	−1	0

Now each row of the table corresponds to a weighing, namely, put the +1s on the left and −1s on the right—and there is it. Perform the weighings, write down the outcome, and read off the guilty coin from the table. (The capital letters of the tables differ by a permutation from the ones given earlier, but this clearly makes no difference.)

In some cases, nonbranching solutions are found easily. I have picked a number between 1 and 8 and you must guess it by asking three yes-or-no questions. Of course, everyone uses the branching, or "interactive," strategy of successive bisecting, but one need not do this. Why not just ask in advance these three questions, in any order. Is the number in the set {1, 2, 3, 4}? in the set {1, 2, 5, 6}? in the set {1, 3, 5, 7}? Of course, this would not work if the allowable question had to be of the form Is the number greater than *x*? The example shows what is going on generally. There is a set of possible states, and one wants to learn the true state. Every question, or weighing, or "experiment" gives a partition of this set. One defines the intersection of k partitions in the obvious way as the partition formed by all intersections of sets of the k partitions. Then the true state can be learned in n experiments without branching if and only if one can find n partitions whose intersection is the partition by singletons.

For our next examples we have chosen two well-known problems involving four coins, each of which is a good or counterfeit one. Now we do not know how many of each there are, and we are required to find exactly which are which.

Three Weighings on a True Scale

For any chosen subset, our "scale" will tell us exactly how many of them are good coins (not which ones but how many of them).

Our solution, again unbranched, is obtained by taking as our first subset coins 1 and 2; second subset, coins 2 and 3; third subset, coins 1, 3, and 4.

So these are our three "weighings," and the method of determination from the three answers, call them a, b, and c, is quite charming. First we add these three and get $a + b + c$, which counts coin 1 twice, coin 2 twice, coin 3 twice, but coin 4 only once. Thus, by reducing $a + b + c$ modulo 2, we obtain the nature of coin 4 itself. Then subtracting this from c, we obtain the balanced system for coins 1 and 2, coins 2 and 3, coins 1 and 3. This then determines the nature of the first three coins.

Seven Questions to a Liar

Back again to our four coins, but this time we may ask any yes-or-no questions about them. The person we are asking, however, is permitted to lie in response to (at most) one of these questions. As a result we must ask more than the obvious four questions we would need with a truth-teller. Indeed, using branching, the correct number of questions needed from this liar is seven.

Once more our purpose is to achieve this same determination with seven non-branching questions. Our first four questions are simply, Is 1 a good coin? Is 2 a good coin? Is 3 a good coin? Is 4 a good coin? The next three questions may seem to use branching, but in fact, the questions do not, though the *answers* do.

> Question 5: Were your answers to questions 1, 2, and 3 all correct?
> Question 6: Were your answers to questions 2, 3, and 4 all correct?
> Question 7: Were your answers to questions 1, 2, and 4 all correct?

The determinations are obtained then as follows: If at most one of questions 5, 6, and 7 was answered NO, then all four of the first answers were true and the coins are determined. If answers to 5, 6, and 7 were YES, NO, NO, then the answer to question 4 was a lie. So negate that one and obtain the correct determination. Similarly, if the answers are NO, YES, NO, just negate the answer in question 1. If the answers are NO, NO, YES, negate answer 2. And finally, if they are NO, NO, NO, then negate answer 3.

More on Squaring Squares and Rectangles

David Gale

Squaring a square or rectangle means tiling (partitioning) the rectangle or square as a union of subsquares. For rectangles with commensurable sides, there are the trivial tilings [letting h be the height and w the width, if $h = (p/q)w$, then tile with a $p \times q$ array of squares of size $h/p = w/q$]. Of interest, however, are *perfect tilings*, where no two squares are the same size. A second natural restriction is to require that no subset of squares form a subrectangle. Such tilings are called *simple*. Here is a brief and very incomplete chronology of some of the work on this problem.

1903. Dehn proves that if a rectangle is tiled by squares, the sizes of all squares must be commensurable. (By the size of a square, we mean the length of its sides.)
1925. Moron finds a perfect tiling of a rectangle by nine squares. (We call the number of squares the *order* of the tiling.)
1939. R. Sprague publishes the first example of a squared square. It has order 55.
1940. Brooks, Smith, Stone, and Tutte prove that no perfect tiling of a rectangle has order less that 9, and there are exactly two tilings of this order.
1948. Willcocks finds a perfect (but not simple) squared square of order 24.
1960. C. J. Bouwkamp and associates find 4094 simple rectangle tilings (3663 perfect, 431 imperfect) of order less than 16, including 2609 perfect ones of order 15.
1962. A. W. J. Duijvestijn proves that there are no simple perfect squared squares of order less than 21.
1978. Duijvestijn finds a simple perfect squared square of order 21 and shows that it is the only one of than order.
1992. Bouwkamp and Duijvestijn publish an illustrated "catalogue" of all simple perfect squared squares (up to obvious symmetries) of orders 21 through 25, containing 207 tilings—one of order 21, 8 of order 22, 12 of order 23, 26 of order 24, and 160 of order 25.

If one takes the square being tiled as the unit square, then from Dehn's result, the size of all the subsquares will be rational. So one can normalize by a uniform stretching, making all sides integers that are relatively prime. The size of the big square is then called the "reduced size" of the tiling. Of the 160 tilings of order 25, the smallest has reduced size 147, the largest 661.

Even 30 years ago one knew more than 2600 perfect tilings of rectangles by 15 squares, so one would expect the number of tilings of rectangles of order 25 to be very large indeed, especially if one does not require the tilings to be simple or perfect. Bouwkamp says that there are about 5,000,000 perfect simple squared rectangles to every such square (for order greater than 20)! Can one even be certain that as the order gets larger the number of tilings remains finite? The answer is yes, but the proof is not completely obvious. However, it might fit nicely as a nonroutine application in an undergraduate course in linear algebra. Further, the proof gives Dehn's theorem as a by-product.

Extend the horizontal sides of all the tiles. The regions between consecutive lines will be called *strips*. In this figure, there are nine squares and five strips.

The original example of Moron.

If there are m strips and n squares, we construct the $m \times n$ *horizontal intersection matrix* $\mathbf{A}$ of the tiling by the following rule: a_{ij} is 1 if strip i meets the interior of square j and is 0 otherwise. The intersection matrix for the figure shown is

1	2	3	4	5	6	7	8	9
1	1	1	0	0	0	0	0	0
1	0	1	1	1	0	0	0	0
1	0	0	0	1	1	0	0	0
0	0	0	0	0	1	1	1	0
0	0	0	0	0	0	1	0	1

Let $\mathbf{x} = (x_1, x_2, \ldots, x_n)$ be the vector whose components are the sizes of the n tiles, and let $\mathbf{y} = (y_1, y_2, \ldots, y_m)$ be the heights of the m strips. Let $\mathbf{1}$ be the m-vector all of whose entries are 1s.

Theorem. *The vectors $\mathbf{x}$ and $\mathbf{y}$ are determined, up to multiplication by a constant, by the matrix $\mathbf{A}$ and are the unique solutions of the equations*

$$(1) \qquad\qquad \mathbf{A}\mathbf{A}^{\mathsf{T}}\mathbf{y} = \mathbf{1} \quad \text{and} \quad \mathbf{x} = \mathbf{A}^{\mathsf{T}}\mathbf{Y}.$$

The height, hence width, of each square is the sum of the heights of the strips that meet it, so by the definition of $\mathbf{A}$, $\mathbf{x} = \mathbf{A}^{\mathsf{T}}\mathbf{y}$. Also, the sum of the widths of all squares that meet a given strip must be the width w of the rectangle, which we may take to be 1. So, again by the definition of $\mathbf{A}$, we must have $\mathbf{A}\mathbf{x} = \mathbf{A}\mathbf{A}^{\mathsf{T}}\mathbf{y} = \mathbf{1}$.

It remains to be shown that $\mathbf{A}\mathbf{A}^{\mathsf{T}}$ is nonsingular, so that $\mathbf{y}$ and $\mathbf{x}$ are unique. We show first that $\mathbf{A}$ has rank m. To see this, note that for any two consecutive strips, say k and $(k + 1)$ reading downward, there is a square that intersects the first but not the second. The corresponding column of $\mathbf{A}$ will have a 1 in row k and 0s in all subsequent rows. Choosing such columns for all k (columns 2, 3, 5, and 7 in the earlier example), we get an upper triangular square submatrix with 1s on the diagonal, which is, therefore, nonsingular. It is then a familiar fact that $\mathbf{AA}^T$ is nonsingular. (This is, indeed, exactly the result used in obtaining the formula for least squares approximations; see, for example, Strang, *Linear Algebra and its Applications,* p. 102. Here it comes up in quite a different context, although this one also involves squares!)

The finiteness now follows. Given the number of tiles, there are only finitely many possible incidence matrices, and given the sizes of the squares, there is only a finite number of possible tilings.

Dehn's theorem follows. The matrix $\mathbf{A}$ is rational, so the unique solution of (1) must be rational.

Because the theorem is proposed for a linear algebra text, the next thing would be

Exercise: Use (1) to find the sizes of the nine squares in Moron's example.

The Two Cultures

The following is from a TV interview with the poet Robert Frost.

> We rise out of disorder into order, and the poems that I make are little bits of order. If I make a basket or a piece of pottery and a vase or something. . . . If you suffer any sense of confusion in life the best thing you can do is make little forms, blow cigarette smoke rings (even those have form, you know). . . .

To this list a mathematician might add, "*or prove a theorem.*"

chapter 13

The Return of the Ant
and the Jeep

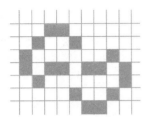

FURTHER ANT-ICS

Jim Propp

The industrious ant (see Chapter 10) has some sisters and cousins that are even more interesting. These generalized ants move from cell to cell in an infinite square grid. At any given moment, each cell in the grid is in some particular state. These states will be numbered 0 through $(n - 1)$, where n is the number of allowed states. When an ant passes through a cell that is in state k, the state of the cell changes for k to $(k + 1)$ modulo n, and the ant then leaves the cell, making either a left or a right turn relative to the direction it was traversing when it arrived at the cell. The ant is not free to choose which way it will go (right versus left). It must proceed in accordance with a *rule-string* of length n that is fixed for all time. The rule-string consists of n bits, numbered 0 through $n - 1$. When the ant is leaving a cell whose state used to be k (and is now $(k + 1)$ modulo n), it turns right if r_k is **1** and left if r_k is **0**, where r_k is the kth bit of the rule-string. This generalization of the ant seems to have been first considered by Greg Turk [1] and, independently, by

Bunimovich and Troubetzkoy [2], who were building upon earlier work of E. G. D. Cohen [3].

We may as well focus our attention on rule-strings that begin with a **1,** which complement all of the bits in a rule-string that simply interchanges "left" and "right" and thus gives a mirror-image universe not essentially different from its twin. We will interpret a rule-string that starts with a **1** as the base-2 representation of a natural number. For instance, Langton's original rule will be called rule **10,** or rule 2. It is easy to see that rule **1** is trivial and causes an ant to travel endlessly around a 2 × 2 square. For the same reason, rules **11, 111, 1111,** and so forth are also trivial.

These generalized ants constitute a special case of the "tur-mites" studied by Greg Turk and others, which were described in A. K. Dewdney's article "Two-dimensional Turing machines and tur-mites make tracks on a plane" (see [1]). Tur-mites are so general that they include Turing machines as a special case. Consequently, it is nearly impossible to prove any general theorems about tur-mite behavior. For the ants, however, one can prove at least one result:

Theorem. *An ant's trajectory is always unbounded, provided that the rule-string contains at least one* **0** *and at least one* **1.**

The proof is the same as the one given in Chapter 10.

There are many questions we can ask about generalized ants. The one I will discuss here is, What happens when you start an ant in a universe in which all cells are originally in state 0? We will assume that the ant's initial heading is southward.

Figure 13.1 shows the state of the universe after an ant with rule-string **10** has been wandering around for 11,000 steps in a universe in which all cells were initially in state 0. State 0 is drawn in white, state 1 in black. We can see a highway forming in a northwesterly direction. (This highway has the same structure as the one shown in Figure 5 of Chapter 10; it is merely rotated by 90°. The same picture appears in Dewdney's article.)

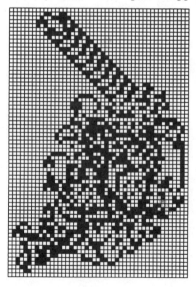

FIGURE 13.1.

To handle ants with rule-strings of length $n > 2$, let white represent state 0, black represent state $(n - 1)$, and intervening shades of gray represent the intermediate states.

FIGURE 13.2.

Ant 4, like ant 2, starts out by creating various symmetrical patterns (such as the one shown in Figure 13.2, which comes into existence at the 236th step); these patterns possess bilateral symmetry, unlike the patterns created by ant 2, which have 180° rotational symmetry. The ant then stops behaving symmetrically and creates a seemingly chaotic jumble, as shown in Figure 13.3 (step 100,000). Is some sort of highway eventually formed? I don't know. I've tracked it for over 150,000,000 steps without seeing any clear pattern.

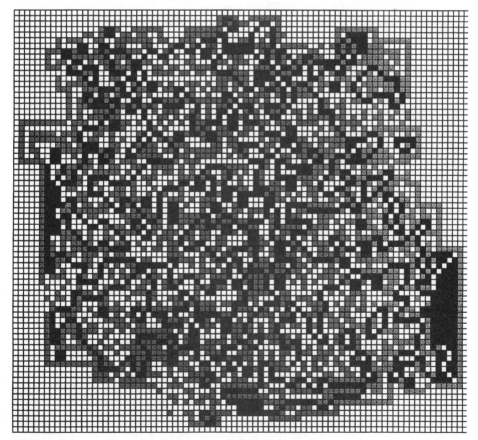

FIGURE 13.3.

Ant 5 is even more like ant 2, in that it favors twofold rotational symmetry; its crowning accomplishment is the pattern shown in Figure 13.4, created after 616 steps. After that, however, the pattern breaks, and after 150,000,000 steps, one sees a configuration with no signs of any reemerging structure. Nevertheless, some surprising statistical patterns appear. For instance, I looked at a central 21×21 square in the middle of the configuration and found only 79 cells in state 0, compared with 190 1s and 172 2s. What is the cause of this fluctuation?

FIGURE 13.4.

In contrast to ants 4 and 5, ant 6 is very tame—even tamer than ant 2. After just 150 steps, one can see a highway forming to the southwest (Fig. 13.5). Unlike the highway formed by ant 2, which has a "period" of 104 (that is, it takes 104 time steps for the ant to build each successive piece of highway), the highway formed by ant 6 has a period of only 18. Moreover, experiments show that even if one modifies the initial state of the universe by sprinkling a few 1s and 2s among the 0s, highways tend to form extremely quickly.

FIGURE 13.5.

Ants 8 through 14 reveal new phenomena. Ants 8 and 14 are the fully "chaotic"; the patterns they build, like those built by ants 4 and 5, show no signs of global structure, though each one has distinctive local motifs, especially along the boundary of the growing cloud of chaos. Ant 10 (with rule-string **1010**) is just an elaborated version of ant 2 (with rule-string **10**); more generally, a rule-string that consists of two or more repetitions of a shorter rule-string will lead to the same behavior as with the shorter string. Ant 13 starts out chaotically, but after roughly 250,000 steps, it starts building a highway of period 388. Ant 14 is a curious hybrid of ants 2 and 6; like ant 2, it builds a highway of period 52, but the highway looks very much like the one built by ant 6. Are these two similarities (one numerical and one pictorial) merely coincidental?

Ants 9 and 12 are the truly surprising ones. In each case, the patterns gets ever larger, but without ever getting too far away from bilateral symmetry! More specifically, one finds that the ant makes frequent visits to the cell it started from, and when it does, the total configuration quite often has bilateral symmetry at the instant that the ant arrives at the starting cell. This phenomenon was first noted by Greg Turk. Figure 13.6 shows ant 12 after 16,464 steps. (This portrayal of M.I.T.'s mascot, the beaver, is offered in appreciation for the use of M.I.T.'s facilities.) Figure 13.7 shows the same ant after 186,848 steps, and Figure 13.8 shows ant 9 after 38,836 steps. (For a picture of this configuration at a later stage, see page 182 in [1].) Greg Turk informs me that Bernd Rümmler has proved that ant 12 builds ever-larger bilaterally symmetric patterns for all time [see Chapter 18].

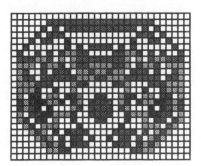

FIGURE 13.6.

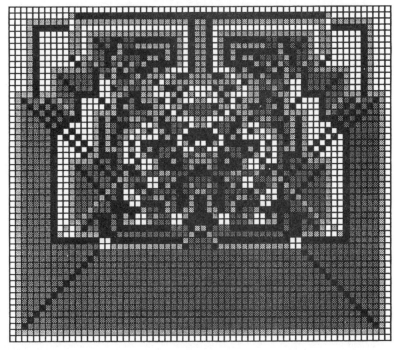

FIGURE 13.7.

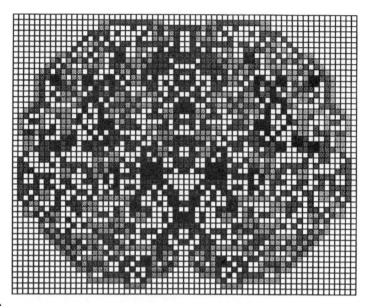

FIGURE 13.8.

If we look at five-bit rule-strings (corresponding to universes with five states), the major new behavior we find is exemplified by ant 27, which builds an ever increasing spiral, as shown n Figure 13.9. We do not, however, find any ants that build ever larger bilaterally symmetrical patterns, like ants 9 and 12. To find more ants of this sort, we have to move on to rule-strings of length 6. Here we encounter another mystery: the rule-strings of length 6 that lead to bilaterally symmetrical patterns are 33, 39, 48, 51, 57, and 60. Note that all these numbers are divisible by three! Surely this cannot be an accident.

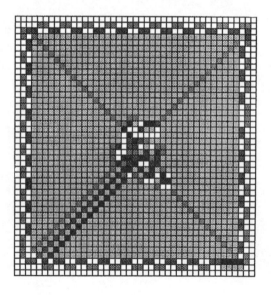

FIGURE 13.9.

The study of generalized ants could easily become a game between two teams. On one team are the "theorists," who will try to develop general rules describing what long-term behaviors are possible and which behaviors will occur given a particular ant rule and a particular initial state of the universe. On the other team are the "engineers," who will try to devise patterns in the ant universe that can be used as building blocks for a general-purpose computer. If the engineers succeed in this, then it will follow that the theorists' goal is in some sense unreachable, just as in the case of Conway's game of Life (see Chapter 25 of [4]). It might even be the case that a simple question like Does ant 4 ever build a highway? is unprovable in your favorite axiomatic basis for mathematics (ZFC or whatever).

Readers are encouraged to play around with ants on their own and draw their own conclusions.

Thanks to L. A. Bunimovich, E. G. D. Cohen, X. P. Kong, Chris Langton, Bruce Smith, S. E. Troubetzkoy, Greg Turk, and Fei Wang, on whose work this chapter draws.

References

1. A. K. Dewdney, Computer recreations, *Scientific American* (September 1989), pp. 180–183; follow-up (March 1990), p. 121.
2. L. A. Bunimovich and S. E. Troubetzkoy, "Rotators, periodicity, and absence of diffusion in cyclic cellular automata"; *Journal of Statistical Physics* 74 (January 1994).
3. E. G. D. Cohen, New types of diffusion in lattice gas cellular automata, in "Microscopic Simulations of Complex Hydrodynamic Phenomena," M. Mareschal and B. L. Holian ed., Plenum Press, 1992.
4. E. R. Berlekamp, J. H. Conway and R. K. Guy, "Winning Ways," Academic Press, 1982.

The Return of the Jeep

No doubt many readers know about the so-called jeep problem, but in case some of you may have forgotten—the problem is to get a jeep across a desert. The difficulty is that the jeep can carry only enough fuel to go part of the way. The solution is to allow the jeep to make preliminary forays into the desert to deposit various amounts of fuel at depots along the way, thus allowing it to refuel as needed during the final trip. Can the jeep cross an arbitrarily long desert in this way, and if so, how can this be done so as to consume the minimum amount of fuel?

A complete solution to this problem was given by N. J. Fine in 1947. Since then, numerous variants of the problem have been treated. In particular, your obedient servant noted in 1970 that if the problem was to send not one but *n* jeeps across, then the minimum cost was strictly less than *n* times the cost for a single jeep (*American Math. Monthly* agreed to publish the result, and even allowed the article to be subtitled "Jeeper by the Dozen").[†] In that article, I listed as unsolved the seemingly natural problem of the most economical way for a jeep to make a trip across the desert and back, assuming that fuel was available at both ends. I'm happy to announce that now, 23 years later, this problem

---

†See Appendix 2.

has been solved by Alan Hausrath, of Boise State University, and Bradley Jackson, John Mitchem, and Edward Schmeichel, of San Jose State. Their solution is quite elegant, and I will attempt here to describe it qualitatively. Before doing so, however, I will look once more at the original jeep problem, whose solution, as others have noted, can be made quite plausible by some commonsense arguments.

First, it turns out to be more convenient to consider the clearly equivalent problem of calculating the maximum distance a jeep can go on x tankloads of fuel, and we shall adhere to this formulation from now on.

In thinking about the problem, I suspect most people would picture the jeep scurrying back and forth between the starting point and the various depots, depositing or picking up fuel as it goes. Suppose, however, that each time the jeep returned to the home base it was replaced by a new jeep for the next outward foray. This surely would not affect the problem. Further, the new jeep could be driven by a new chauffeur. But if that were the case, one could save a lot of time, because there would be no reason for each jeep to wait for the return of the preceding one before setting out. They could all leave the starting line and travel together as a convoy, the purpose of all but one of the jeeps being to refuel the others. Figure 13.10 is a schematic bird's-eye view of a four-jeep convoy that has set out from S in the direction of F.

We choose as the unit of distance the distance a jeep can travel on a tankload of fuel. In the figure, the plaid (or superjeep) J^* is supposed to make the trip. The others are the refuelers. The refuelers must return to the starting line, which is equivalent to assuming that they consume twice as much fuel per unit distance as the superjeep. The shaded portion of the jeeps represents the fuel remaining in their tanks after they have gone a distance x. So the unshaded portion represents x units of fuel for the superjeep and $2x$ for the others. At the time shown in Figure 13.10, Jeep #1 is about to give up all of its fuel, $1 - 2x$ units, to exactly fill up the other three. So $1 - 2x = 5x$ or $x = 1/7$. The four-

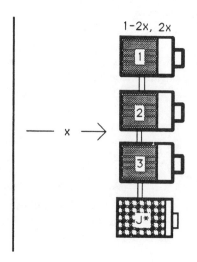

1-2x, 2x

S   F

FIGURE 13.10.

convoy problem has now become a three-convoy problem. It is now easy to see that using this method, an n-jeep convoy can travel a distance

$$1 + \frac{1}{3} + \frac{1}{5} + \cdots + \frac{1}{2n-1}.$$

In case the given amount of fuel is not an integer, let f be the fractional part, and add an extra refueling jeep to the convoy to carry the extra f units. A calculation like the one above shows that one gains an additional $f/(2n+1)$ units of distance. For future use, define $D(x)$ as the maximum distance a jeep can go on x units of fuel. Then

(*) $$D(x) = 1 + \frac{1}{3} + \cdots + \frac{1}{2n-1} + \frac{f}{2n+1},$$

where f is the fractional part of x and $n = x - f$.

The problem in which the jeep must cross the desert and return is called the *round-trip problem*, and here the formula is even simpler, $1/2 + 1/4 + \cdots + 1/(2n)$, as follows easily from the convoy formulation. Likewise, for the k- (or dozen) jeep problem, the same model shows that on n tankloads of fuel, $n > k$, the k jeeps can go distance $1 + 1/(k+2) + \cdots + 1/(2n-k)$. Of course, in all these cases some further argument is needed to show that these formulas actually give the optimal distance.

Before going on to describe the solution of what I will call the *two-way jeep problem* (round-trip with fuel available at both ends), let me digress for a moment to consider a more general convoy problem. Instead of assuming that all jeeps are alike, we consider different kinds of jeeps. A jeep is characterized by two numbers, its *capacity* C = the number of liters it can carry, and its *fuel efficiency* E = the number of kilometers it can travel on a liter of fuel. Problem: Given n jeeps, each with its own C and E, what is the longest desert that can be crossed?

I believe that there is a literature on this problem (the context is usually rockets and interstellar space rather than jeeps and deserts), but as far as I know, the general problem remains unsolved, meaning there is no known "good algorithm" for finding the optimum. Presumably there is some refueling sequence that achieves the optimum, but what is it? One way to visualize the situation is again given by Figure 13.10, but this time, imagine that the jeeps are rockets and the figure is in a vertical rather than horizontal plane. Then the refueling can take place continuously. Gravity causes the fuel to run out of the top rocket, keeping the lower ones full at all times. When the top tank is empty, the top rocket is abandoned, the others continue on, and the process repeats.

It is conceivable that this problem is NP-hard. As with many optimization problems, the difficulty is that there is no easy way to recognize when a given solution is optimal. How does one deal with this type of situation? There is one general method that sometimes works. One shows that any optimum solution must satisfy certain conditions (e.g., at an interior maximum of a differentiable function, the derivative must vanish). If one is lucky, one finds enough such conditions so that a unique object satisfies them, which must, therefore, be the optimum. This, in fact, is what the authors do for the case of the two-way jeep problem, which I now describe.

We first consider the problem of the longest desert that can be crossed if there are f units of fuel at S and g units at F. Some more common sense: If $g \geq f$, then clearly the best one can do is $D(f)$ [given by (*)], and one uses S-fuel for the outward trip and F-fuel for

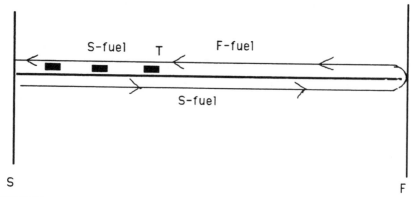

FIGURE 13.11.

the return. On the other hand, this procedure is clearly nonoptimal if $g < f$, for then one could do no better than $D(g)$, and some of the S-fuel would remain unused. There must, therefore, be some way of setting up depots of S-fuel on the outward trip that can be used on the return; so imagine that such depots have been set up, and suppose the one farthest from S is located at a point T, as shown in Figure 13.11.

The authors now prove that there are optimal solutions with the following highly plausible properties. Obviously, on the outward trip only S-fuel can be used, but it is shown that there is an optimal solution such that (i) on the return trip only F-fuel is used in getting from F back to T, and (ii) only S-fuel is used for getting from T back to S (as indicated in Figure 13.11). Intuitively, it surely seems wasteful to bring S-fuel to F (coals to Newcastle) or to bring F-fuel further back than T (if you needed the extra fuel, you should have left it there on the outward trip).

But it turns out that these conditions actually determine the solution, namely, the distance d_1 from T to F must be $D(g)$, because from (ii), all of g and nothing else must be used up in going from F to T. Therefore, on the outward trip the jeep must transport g units of fuel to T. The distance d_2 from S to T must be the optimal distance a round-trip jeep can go on f units of fuel if it is required to deliver g units at its destination; this is just a mild modification of the original round-trip jeep problem and is easily solved by the convoy method. Then $d = d_1 + d_2$ is the distance we seek. Thus, the solution for the two-way problem is reduced to patching together the solutions of the two original jeep problems.

The authors now treat another two-way problem: given x units of fuel, how should one choose f and g optimally, where $f + g = x$? For $2 \geq x$, the best one can do is put half the fuel at each end. For $x > 2$, the solution is, in general, not unique, but there is always a solution in which g is an integer given by $g = [((x+1)/2)^{1/2}]$, much less than half of the fuel for large values of x, although the optimal distance traveled differs from the distance when half the fuel is at each end by no more than $1 + \ln 2$ (independent of x). Further, the number of intermediate depots needed for the return trip is $[x] - 2$. Thus, Figure 13.11 represents the solution in which there are between four and five tankloads of fuel available.

Of course, the arguments proving the optimality of the authors' algorithm are the main content of their article. All we have done here is to describe what that algorithm is.

chapter 14

Go

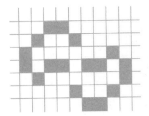

THE PROBLEMS

Chapter 11 of this book is devoted to certain games, and we remarked that although game theory has become an extensive branch of both pure and applied mathematics, the set of games studied in the mathematical literature, is with very few exceptions, disjoint from the set of games people actually play. A notable exception is the analysis by Paterson and Zwick of the children's game of *Concentration*, where the authors actually found the rules for optimal play; that is, they solved the game.

This chapter is also devoted to a game people play, in fact, one of the most played and important games in human history, the game of go; we report on some recent and continuing work of Elwyn Berlekamp and his associates, David Wolf, David Moews, Yonghoan Kim, and Raymond Chen. We hasten to remark, however, that, unlike Paterson and Zwick, Berlekamp has not solved go. In fact, it is not clear how much help, if any, his work offers to people or even computers in improving their game. What he has done is to devise a set of go problems that computer program can solve but that

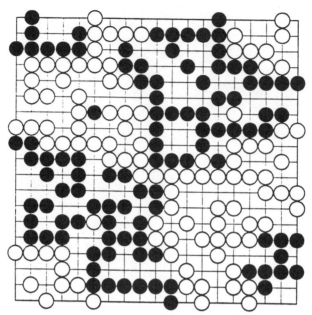

FIGURE 14.1. White to play and win.

no professional player has succeeded in solving. The board set up in Figure 14.1 is an example.

Taken by itself this may not be so impressive. Indeed, there are instances of chess problems devised with the aid of computers that one would not expect even Garry Kasparov to solve. An example is given in Figure 14.2.

FIGURE 14.2. Black to move, white to win (promote, capture, or mate) in 109 moves.

It seems doubtful that any human player would recognize this position as a win for white, since black can prolong the agony for so many moves. However, aside from being a curiosity, this problem is of little interest. In contrast, what makes Berlekamp's work notable is that to solve his problems, one must apply a vast, profound, and beautiful mathematical theory, called "combinatorial game theory" by its inventors, Berlekamp, Conway, and Guy. There are only a handful of mathematicians in the world who know this theory, which is expounded in the authors' 850-page, two-volume book *Winning Ways*. Before describing the applications to go, I will attempt to give a thumbnail sketch of what this theory is about, that is, to summarize 850 pages in a few paragraphs—which obviously can't be done, but anyway, here goes. (In fairness it should be said that only about 150 pages are devoted to the theory; the bulk of the book is concerned with applications to specific games.)

COMBINATORIAL GAME THEORY

The theory treats only win-lose games, between Ms. Left and Mr. Right, in which the person who moves last is the winner. Although this seems rather special, many games can be put in this form by making simple changes in the rules. Such games are conveniently represented as trees like the one in Figure 14.3.

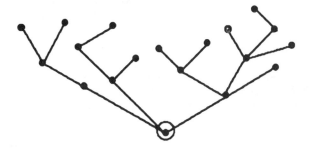

FIGURE 14.3.

The "button" is originally at the root, as shown. Depending on whose turn it is, the button is pushed along some edge either left or right, and the winner is the one who pushes it onto a terminal vertex.

A simple but highly nontrivial example of such a game is *domineering*, illustrated in Figure 14.4.

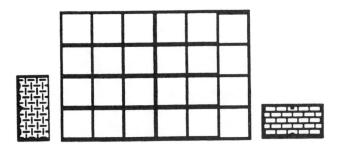

FIGURE 14.4.

The rules are simple. Left has the vertical dominoes and right the horizontal; they take turns placing them on the grid—horizontally rsp. vertically so as to cover exactly two squares. Dominoes are not allowed to overlap. A player loses if she is unable to move—and that's it. I chose 4 × 6 Domineering as an example because at the time of this writing this game is still unsolved.

Returning to the general theory, all games belong to one of four types:

1. a win for Left (regardless of who moves first);
2. a win for Right (regardless of who moves first);
3. a win for the player who moves first;
4. a win for the player who moves second.

Each of the four possibilities is illustrated by the Domineering games in Figure 14.5.

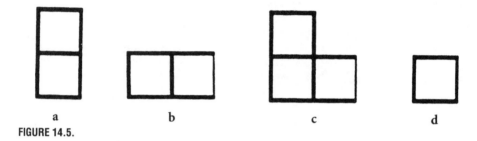

    a   **b**   **c**   **d**
FIGURE 14.5.

We call Figure 14.5d a second-player win because the first player is unable to play on his first move. A less trivial example of a second-player win is the board presented in Figure 14.6, shown with its tree. The reader will easily verify that this is a second-player win.

    a   **b**
FIGURE 14.6.

The first observation is that there is a natural notion of *addition* on the set of all games, namely, the sum of two games G and H is defined as the two games played simultaneously. Thus, take (Checkers) + (Tic-Tac-Toe) (suitably modified to make them win–lose). Left, say, plays the black pieces and makes X's, while Right plays red and makes O's. A player on his turn may play in either game.

Next, the negative, $-G$, of a game G is defined as G with the roles of the players interchanged (reflect the tree about a vertical axis).

Now comes the key observation: We define an equivalence relation $\sim$ on G by writing $G \sim H$ if $G - H$ is a second-player win. Observe that $\sim$ is reflexive: The second player can win $G - G$ by playing "copycat": whenever the first player plays in G $(-G)$, she plays the same move in $-G$ (G). From now on, instead of saying G is a second-player win, we will write $G \sim 0$. One must now show the following:

1. If $G \sim 0$ and $H \sim 0$, then $G + H \sim 0$ (Do you see it?).
2. If $G \sim 0$, then $G + H \sim H$ for all H, because $(G + H) - H = G + (H - H) \sim 0$, by reflexivity and (1).
3. Symmetry: $G \sim H$ if and only if $H \sim G$ [because $- (G - H) \sim H - G$, etc.].
4. Transitivity: If $G - H \sim 0$ and $H - K \sim 0$, then $G - H + H - K \sim 0$, so $G \sim K$.

So (you saw it coming) the set of equivalence classes forms an abelian group whose 0 element is the set of all second-player wins. We will call this group Γ, and it is the object of study in combinatorial game theory. From now on, G denotes an equivalence class, so $\sim$ can be replaced by $=$. To get a feeling for Γ, consider the Domineering games in Figure 14.7.

Both games are obviously wins for Left, but Figure 14.7a is the *same* game as Figure 14.5a (prove it), whereas Figure 14.7b is not. Figure 14.5a–Figure 14.7b is a win for Left rather than the second player. (If you don't go through these verifications, you're missing half the fun.)

Back to the theory. There is a natural partial order on Γ. Call G positive, $G > 0$, if G is a win for Left, and negative if G is a win for Right. Note that $>$ is well-defined; i.e., games equivalent to a positive game are positive (why?). Obviously the sum of positive games is positive (Left should always respond in whichever game Right plays), so defining $G > H$ if $G - H > 0$ gives a partial ordering of Γ. Every game is thus either positive

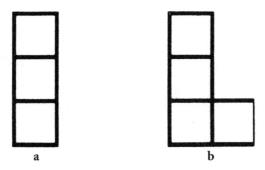

a   b

FIGURE 14.7.

(Fig. 14.5a), negative (Fig. 14.5b), zero (Fig. 14.5d), or none of the above, meaning a first-player win (Fig. 14.5c).

So now that we have a partially ordered abelian group, what do we intend to do with it? At this point the analysis takes a surprising turn. In the usual algebraic approach to group theory, one proceeds to study the structure of the group, its subgroups, representations, and so on, but here the interest is not in structure but in the individual group elements. Indeed, these enter the picture somewhat like characters in a play or novel. They even have names, some of which are numbers, others are Star, Up, Down, Tiny, and Miny (with the symbols $*$, $\uparrow$, $\downarrow$, $+_{on}$, $-_{on}$), and there are dozens more. Further, each of these elements has not only a name and a symbol but also a "picture," namely, its tree. Of course, many different trees correspond to the same game. For example, the tree of Figure 14.6b represents the game 0 whose tree is simply a single vertex. One of the important theorems of the theory states that every game has a unique canonical form, meaning a unique tree with the fewest possible branches. For example, the canonical form for Figure 14.7b is given by Figure 14.8.

FIGURE 14.8.

Γ has many interesting subgroups. Perhaps the simplest is $\{0, *\}$ where $*$ is the game of Figure 14.5c, whose tree is given in Figure 14.9.

You should verify that $* + * = 0$. In fact, any game whose tree is right–left symmetric must clearly have order 2. It turns out that the canonical form of the tree of Figure 14.3, which I constructed just by playing around, is the tree of $*$ (Fig. 14.9), as Berlekamp pointed out to me. Obvious question: Given a game, how does one find its canonical form? Answer: The general problem is NP-hard.

The most important subgroup of Γ is a group called Numbers, and it is isomorphic to the group of dyadic rationals. The tree of n is simply

FIGURE 14.9.

and it is a triviality to show that $m + n = (m + n)$. More interesting is $1/2n$, whose tree is

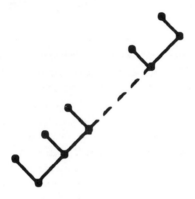

Thus, Figures 14.8 and 14.7b are $1/2$. A nice exercise is to show that $1/2^n + 1/2^n = 1/2^{(n-1)}$. Thus, Numbers add just like numbers. The tree of $3/4$ is

Recall that Γ is partially ordered so one may ask, for example, where does $*$ fit into the ordering. It is an easy exercise to show that $*$ is an "infinitesimal," meaning that it is smaller than any positive Number and greater than any negative Number, but not comparable with 0.

My problem now is deciding when to stop, because there is so much more that could be said. For example, games have properties called *temperature* (every game is either hot, cold, or tepid) and *atomic weight*. There are important homomorphisms from Γ to Numbers called *cooling* and *chilling*; in some cases, *chilling* has an inverse called *warming*; and on and on. It's time, however, to move on toward explaining the connection with go.

SOLUTIONS AND DECOMPOSITION

What does it mean to solve a game? A person not initiated into combinatorial game theory (henceforth to be abbreviated CGT) might suppose it means knowing who the winner is and how to go about winning, that is, to describe the *winning strategy*. The CGT notion of solution is rather different and is at once more and less than the naive notion. A game is said to be *solved* if one knows exactly which element of Γ it is, or, more precisely, if one knows its canonical form. This is called the *value* of the game. (In practice, one is happy if one can express a given game as a sum of games whose values are known. We will take this up in a moment.)

Some examples: The value of 2×6 Domineering is $1 - |_2 - 1$ (don't worry if you don't know what this notation means). It is known that 5×5 Domineering has value 0, so it is a second-player win, though knowing this fact does not mean that one necessarily can obtain the second player's winning strategy without performing a major amount of computation.

Indeed, the Domineering story is quite intriguing. Even games as apparently simple as $2 \times n$ Domineering are far from trivial. Berlekamp has a "formula" for the value when n is odd. It turns out that $2 \times n$ is a second-player win only for $n = 13$. At the time of the writing of *Winning Ways*, 4×4 was unsolved, and a year ago Berlekamp assigned it to a couple of MIT undergrads as a term project. They came up with different answers, neither of which was correct. I asked Elwyn for the correct solution, which he supplied, but this too turned out to have a bug. Finally, using a computer program, David Wolf came up with the correct (we hope) canonical form. The tree turned out to have 52 branches.

But if CGT doesn't tell one how to play games, what good is it? That's a fair question, and the answer is that the theory is extremely useful for a very restricted class of games, namely, those games in Γ that can be expressed as a sum of simple games whose values are known. Such games are said to *decompose*. The prototype of such games is the game of Nim, where one is clearly playing a sum of games, each of which is completely trivial. (The element of Γ corresponding to an n-element Nim pile is called a Nimber and is denoted by $*n$. **Exercise:** Draw the tree for a three-element Nim pile.) Indeed, one way of looking at CGT is as a vast generalization of the theory of Nim.

Now, most games one is likely to run into are not decomposable. For example, the game of Chomp, described in Chapter 11, never decomposes. However, anyone who has played the well-known game of *Dots and Boxes* has encountered an example, par excellence, of decomposition. Recall that in the endgame, the board breaks up into disjoint regions with the property that drawing a line between dots in a region gives the opponent all the boxes in that region. (A whole 43-page chapter of *Winning Ways* is devoted to Dots and Boxes, a Berlekamp specialty, though a complete theory has not yet been found.)

So (finally!) what about go? The positions that occur in the early stages of a game of go surely do not decompose. However, positions that occur in the later stages of the endgame do decompose—both in the technical sense of CGT and in the geographical sense that the board is divided up into separate independent regions—and it then becomes fruitful to apply CGT. Professional go players are very skilled at handling positions involving only a few terms (regions) even when some of these terms are still beyond the reach of CGT analysis. What Berlekamp and Co. have done is to construct certain go positions that do, in fact, decompose and where the values of the games associated with the different regions can be calculated. As examples, all of the go positions in Figures 14.10 and 14.11 below have as values our old friends ½ and *, respectively.

Figure 14.12 shows the decomposition for the problem of Figure 14.1. Gaze and admire!

In this example, white moves and wins by one point but only if he does everything correctly. There is a software package that plays optimally in this game and others in the collected problem set. The software does not look ahead; its computations consist only of rudimentary manipulations of the values, but it is able to play perfectly because it maintains an accurate table of the local values of all regions on the board.

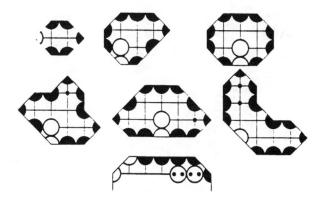

FIGURE 14.10.

FIGURE 14.11.

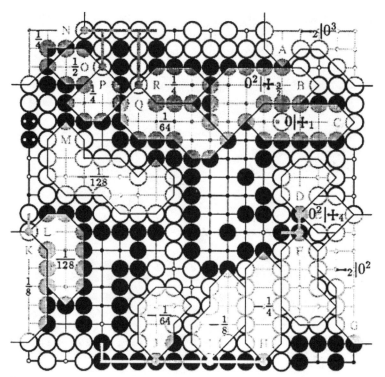

FIGURE 14.12.

LOOKING AHEAD

The natural next question is Where do we go from here? It's clearly fun to stump the experts, but now that one knows it can be done, the point has been made. But there are other reasons to pursue the study. Berlekamp notes that professional go players have formulated various quite sophisticated but rather imprecise principles that they use in playing endgames. He believes that using CGT, some of these "rules of thumb" may be given a precise mathematical formulation. One might think of this enterprise as a very sophisticated analogue to the earliest work in modern game theory, namely, the results of Borel and Von Neumann in giving the complete mathematical theory of bluffing in two-person poker.

SOME AFTERTHOUGHTS

I should perhaps apologize for conducting my education in public in the preceding paragraphs. At the start I knew nothing about combinatorial game theory, and, as should be evident, I have by now barely scratched its surface. Nevertheless, the experience has left me with an overwhelming sense of awe at the unfathomable diversity of mathematics itself. The authors of *Winning Ways* have created a mathematical fairyland. To my knowledge, there is nothing remotely like it anywhere else in the discipline. With this in mind, I can't resist concluding with a bit of speculation.

One often hears people say that all of mathematics is really one, and as the subject develops we will ultimately see clearly how everything fits in with everything else. The whole Bourbaki enterprise, it seems to me, was predicated on some such unspoken assumption. Those authors seem to have been trying to help us to find the "right way of looking at things." But I wonder if this belief in ultimate unity may not be just wishful thinking. Certainly, it would be nice if from a few guiding principles, we could gain an understanding of everything. Unfortunately, there is nothing in the evidence on hand to suggest that this will ever be the case. How much unity will there ever be between, say, the theory of transfinite cardinals, on the one hand, and numerical solutions of partial differential equations on the other?

One hears about unity versus diversity in sciences other than mathematics. The physicists talk about a Grand Unified Theory for which they are searching. Perhaps the projected Super Collider will provide some clues on this, but if experience is any guide, this great piece of hardware may end up creating more puzzles than it solves, in which case I suppose we go on and build the Super Duper Collider.

My own hunch is that mathematics (perhaps physics too) is not going to unify. It's just not in the nature of the beast. It may even be that Bourbaki-like attempts in the end do more harm than good by forcing people's thinking into prescribed patterns. Combinatorial game theory is, I expect, just one example of what unfettered mathematical imagination is capable of creating. There will surely be many others, and I suspect that mathematics, far from becoming unified, will continue to diversify in totally unpredictable ways. This may seem a somewhat terrifying thought, but if, as I believe, this is the kind of universe we live in, then we might as well face up to it.

THEOREMS EVERYWHERE

Shakespeare writes of finding "tongues in trees, books in running brooks, sermons in stones, and good in everything." If he had been mathematically inclined he might also have found theorems somewhere too, perhaps in clouds. Indeed, I expect people who are "mathematically aware" often bump into theorems in unexpected places.

As an example, I recently recalled a puzzle-game which used to appear, perhaps still does, in children's magazines. You are challenged to get, say, from SHIP to DOCK in the fewest possible steps, as illustrated by the following possible solution.

SHIP, SHOP, CHOP, COOP, COOK, COCK, DOCK.

(The fact that this game involves four-letter words has suggested the possibility of more ribald versions, for people who have nothing better to do with their time.)

Here then is the

Ship–Dock Theorem. *In any solution of the problem, there must be a word at least two of whose letters are vowels.*

I am, of course, not proposing this as a challenging problem for the readers of this magazine, but it may be useful in other ways. As already mentioned, it is an example of a theorem that comes up in "everyday life." Further, the result could conceivably be considered to be applied mathematics, being useful in solving the ship–dock problem: it suggests starting in the middle with the double-vowel word rather than at either end. Also it provides another illustration (along with the bridges of Königsberg, Pappus's Theorem, etc.) to convince people that mathematics is not the study of numbers, as many of them believe.

Most interesting to me, however, are possible pedagogical uses of the result. Obviously, no mathematical background is needed to understand the problem. In limited experiments I have tried stating the result and asking people to explain why it must be so (I try to avoid using the word "proof" which seems to induce instant panic in some people). The results have been varied, from (a) near hostility, "Look, I stopped taking tests when I got out of school," to (b) embarrassment, "Let's change the subject," to (c) a glimmer of light, "It's obvious because the vowels in Ship and Dock are in different positions" (you're on the right track), to (d) "Well, I can see it all right but I can't explain it" (where have I heard that before?), to (e) "It's because every word must contain at least one vowel" (good, you're almost home!), to (f) the complete argument, and, better still, (g) the pleasure from seeing how the logical pieces fall together. Of course it's (g) that keeps us going and makes the practice of mathematics such a rewarding profession. As teachers, part of our job is to move our students away from case (a) toward case (g). Unfortunately, so far as I know, no one has yet figured out how to do this. Perhaps the best we can do is to go on scanning the clouds for more theorems.

chapter 15

More Paradoxes.
Knowledge Games

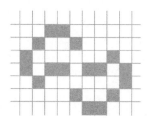

To reintroduce the subject of paradoxes (see Chapter 2) I present the following example.

You and I are assigned numbers a and b which are consecutive positive integers, say $b = a + 1$. Each of us knows our own number but not the other's. Every 10 seconds a beeper sounds, and if either of us knows the other's number, we must announce it immediately after the beep, and the game is over.

This seems like a rather crazy game. What possible help can the intermittent beeping be to the players? Suppose, for example, that my number is 10 and yours is 11. Then we both know that after the first beep neither of us will announce, so when this comes to pass it would seem we are no better off then when we started. Nevertheless,

Theorem. *With perfect play, the person holding n will announce that the opponent's number is $n + 1$ after the nth beep.*

Induction on n: If my number is 1, I will know yours is 2, and I will announce after the first beep. Now suppose that my number is n. If yours is $n - 1$,

then by the induction hypothesis, you will announce after the $(n-1)$st beep, so when you don't do this, I know that your number is $n+1$.

Although the proof is correct, a puzzle still remains. How does a nonannouncement that we both knew would happen change the game? To understand this, suppose my number is 3 and yours is 4. Then we both know in advance that there will be no announcement after the first beep. However, I don't know that you know this because as far as I know, your number could be 2, in which case you would have to allow for the possibility that mine was 1; so an announcement would be forthcoming. The beep, therefore, gives me new information. If our numbers are 4 and 5, then I do know that you know that there will be no announcement after the first beep, but you don't know that I know this, and so on.

This is an example of what have come to be called *common knowledge* games or puzzles. Here is another. In a school for bright children, one morning, in a class of 20 kids, 14 come to school with dirty faces. The teacher says, "We are going to play a game. You can all see each other's faces. This beeper is going to sound every 10 seconds and if you know that your face is dirty, please raise your hand after the beep." The game commences, but after several minutes nothing has happened. So the teacher says, "I see I'm going to have to give you a hint. At least one of you has a dirty face." As every child is able to see at least 13 dirty faces, this would not seem to be big news. Nevertheless, when the game starts again, all the dirty-faced children raise their hands on the 14th beep. The proof, left to the reader, is again by induction, this time on the number of dirty faces.

What new information was gained from the hint? Let's consider the case where only you and I have dirty faces. Before the hint I *know* that you will not announce after the beep. But after the hint there is the possibility that you might; namely, if my face is clean. If there is a third dirty face, then we all know that there will be no announcement after the first beep, but I don't know that you know this—and so on.

The most mathematical variant on this theme is a game invented by John Conway and Michael Paterson [1]. There are N people in a room, and player k has a nonnegative number a_k written on his forehead. In addition, there are N or fewer distinct positive numbers A_k, written on a blackboard, one of which is the sum of the numbers a_k. The numbers need not be integers. Each player sees all numbers except his own. Again we use the 10-second beeper, and the idea is to see how many beeps it takes before someone knows his number, or, what is the same thing, knows which A_k is the true total. The Conway-Paterson theorem asserts that this game always terminates.

Again we have a paradox. Suppose, for example, in a three-player game, all the lower-case a's are 2 an the capital A's are 6, 7, and 8. Then everyone knows that her number is at most 4; hence, her opponents are looking at two numbers whose sum is at most 6. Hence, any of the three A's could be the true total, so there will be no responses after the first beep.

Perhaps the most surprising thing about the Conway–Paterson theorem is that the proof is extremely simple, provided that one goes after it in the right way. Before presenting it, however, let us look at the two-player case with blackboard numbers $A < B$. We will say that a pair of forehead numbers (a, b) is *possible* for the (A, B) game if and only if either $a + b = A$ or $a + b = B$. So the set of possible (A, B) games consists of a pair of diagonal lines in the plane. See Figure 15.1.

Now, the games that terminate after one beep consist exactly of those pairs one of whose members, x, is greater than A, since the player who sees such an x will know that the true total is B. These are the two segments of the B line labeled 1 in Figure 15.1. Hav-

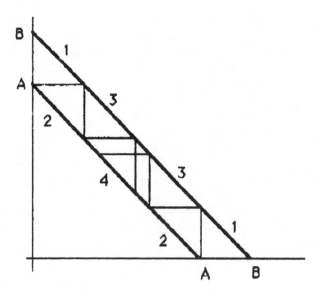

FIGURE 15.1.

ing eliminated all pairs containing a number greater than A, the game ends on the second beep if any pair contains a number less than $B - A$, since a player seeing such a number will know that the true total is A. These are the segments labeled 2 on the A line. In this way, each beep allows one to lop off end segments of one of the two lines. From Figure 15.1 one sees that the entire A line is eliminated after 4 beeps; so, for example, if $a = b = B/2$, then the players will know this after the fourth beep.

Before giving the general proof, we take note of another surprising fact, discovered by Lasry, Morel, and Solimini [2]. The Conway–Paterson game may terminate even when there are more blackboard numbers than players. For example, we will show that the two-player game with blackboard numbers 5, 8, and 15 terminates after at most 10 beeps. Table 15.1 shows the possible pairs of numbers for this game.

TABLE 15.1   Possible Pairs of Numbers

5	8	15
{0, 5} 5	{0, 8} 6	{0, 15} 1
{1, 4} 9	{1, 7} 8	{1, 14}1
{2, 3} 3	{2, 6} 2	{2, 13} 1
	{3, 5} 4	{3, 12} 1
	{4, 4} 10	{4, 11} 1
		{5, 10} 1
		{6, 9} 1
		{7, 8} 7

As usual, the analysis proceeds by first finding all pairs that give a one-beep game, then those that give a two-beep game, and so on. The number after the braces in Table 15.1 indicates, for each pair, the beep on which the game will terminate. First, if either number is greater than 8, then the player who sees it will know that the true total is 15 and will announce it after the first beep. Note that these numbers, 9 through 15, are exactly those that appear in only one of the possible pairs. After deleting these pairs, we see that there is only one pair, {2, 6}, containing a 6, so that if a player sees a 6, she will know that the true total is 8 and announce after the second beep. Eliminating this pair, we find that {2, 3} is the only remaining pair containing 2, so it is eliminated next, and so on. The rule could hardly be simpler. At each stage, look for all numbers contained in only one pair and eliminate those pairs. The 10-beep game is the one in which each player's number is 4.

On the other hand, if the 8 of the above example is changed to 9, we will show that the game cannot terminate. As before, we see that the game will terminate after the first beep if and only if one of the numbers is 10 or higher. However, we are now stuck, because, as shown in Table 15.2, every number occurs in two of the remaining pairs; so no matter what number a player sees, he or she will not be able to draw any conclusions as to the true total.

TABLE 15.2 Possible Pairs of Numbers

5	9	15
{0, 5}	{0, 9}	{0, 15} 1
{1, 4}	{1, 8}	{1, 14} 1
{2, 3}	{2, 7}	{2, 13} 1
	{3, 6}	{3, 12} 1
	{4, 5}	{4, 11} 1
		{5, 10} 1
		{6, 9}
		{7, 8}

In [2], the authors completely analyze two-player, three-number games. Suppose $A_1 < A_2 < A_3$. Then a necessary and sufficient condition for such games to terminate is that there exist integers p and q such that $A_1 < p(A_2 - A_1) + q(A_3 - A_2) < A_3 - A_2$.

With the second example in mind, one can now give the surprisingly simple proof of the Conway–Paterson theorem for the N-player, N-number game. The key idea is to argue by contradiction. Instead of proving that the game must terminate, one proves that it is impossible for it *not* to terminate. As in the above example, for a game to fail to terminate, it must reach a point where after some beep it is impossible to eliminate any more possible N-tuples. When can this happen? The answer is given by a simple theorem about vector spaces.

Let us call a set S of N vectors *ambiguous* if for any member $\mathbf{a}$ in S and any index i there is a vector $\mathbf{a}'$ in S that differs from $\mathbf{a}$ only in the ith coordinate. If a game reaches a point where the remaining set of N-tuples is ambiguous, then no player learns anything from the next beep: There will always be at least two possible values for his own number.

Given a vector $\mathbf{a}$, let $\mathbf{a}_0$ denote the sum of the coordinates of $\mathbf{a}$.

Lemma. *If S is a finite ambiguous set of N-vectors, then the set of sums $\mathbf{a}_0$ must contain at least $(N + 1)$ members.*

Immediate for $N = 1$. Now choose a member $\mathbf{a} = (a_1, \ldots, a_N)$ such that a_1 is a minimum for all $\mathbf{a}$ in S, and let S' be the set of all $(N - 1)$-vectors $\mathbf{x} = (x_2, \ldots, x_N)$ such that $(a_1, x_2, \ldots, x_N)$ is in S. Then S' is also an ambiguous set of vectors. So by the induction hypothesis, the vectors of S' have at least $(N - 1) + 1 = N$ distinct sums. Thus, the vectors $(a_1, x_2, \ldots, x_N)$ have at least N sums. Let $(b_2, \ldots, b_N)$ be a vector of S' whose sum is largest. By the choice of a_1 and the definition of ambiguity, there is $a_1' > a_1$ such that vector $(a_1', b_2, \ldots, b_N)$ is in S, but this yields a sum greater than any of the sums already accounted for; thus, an $(N + 1)$st sum.

We remark that $N + 1$ is "best possible," as illustrated by the example of the set of all N-vectors all of whose coordinates are either a or b. The sum then depends only on the number of b's, which can vary from 0 to N.

As an exercise, the reader may try to show that the three-player game with forehead numbers 2, 2, 2 and blackboard numbers 6, 7, 8 terminates after 15 beeps.

References

1. J. H. Conway and M. S. Paterson, *A Headache-Causing Problem,* in privately published papers presented to H. W. Lenstra on the occasion of the publication of his *Euclidis-che Getallenlichamen.*
2. J. M. Lasry, J. M. Morel, and S. Solimini, On knowledge games, *Revista Mathematica de la Universidad Complutense de Madrid* 2(2/3) (1989).

chapter 16

Triangles and Computers

INTRODUCTION

Chapter 6 took up the subject of computer-aided discoveries in geometry and, in particular, some work of Clark Kimberling on "centers" of triangles, points like the centroid, circumcenter, orthocenter, and so on. Kimberling defined 91 such points and found by numerical explorations that there were (or I should say, appeared to be) an enormous number of collinearities among these points. These empirical results could then be proved, again using computers, but this time using symbolic rather than numerical computation: By expressing the given centers by "trilinear coordinates" as functions of the side lengths, a, b, and c, it became a matter of showing that the appropriate determinant in these symbols vanishes, a task ideally suited to programs like *Mathematica* or *Maple*.

A third possible use of computer technology, one that is especially natural for geometric problems, is, of course, computer graphics. The first of the three examples presented here shows how such computer-generated pictures have led to new and quite striking results in classical Euclidean plane geometry. By contrast, our second example, although even more elementary, deals with a question quite unlike any that are taken up in any treatise on geometry of which I am aware. In the third example, pencil and straightedge replace the computer at the experimental stage, but the "punch line" is once again a result of Kimberling's numerical experiments.

THE DANCE OF THE SIMSON LINES

Given a triangle $\triangle$, let S be its circumscribed circle. From any fourth point P on S drop perpendiculars to the three sides of $\triangle$.

Theorem. *The feet of the three perpendiculars are collinear.*

This locus is called the *Simson line* of $\triangle$ with respect to P (see Fig. 16.1). [According to N. A. Court, *College Geometry* (1925, revised 1952, New York: Barnes & Noble), Robert Simson (1687–1768) is wrongly credited with having discovered this line, which was actually discovered by William Wallace in 1799.]

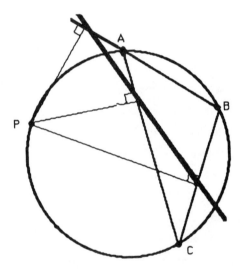

FIGURE 16.1.

Among the Simson lines are the three (extended) altitudes of the triangle and the three extended sides. To get the altitudes, place P at a vertex of $\triangle ABC$. To get, say, side BC, place P at the antipode of A. (Draw the picture and you will see the simple proof of this.)

Next, given an inscribed quadrilateral $ABCD$, four Simson lines are determined, that of $\triangle ABC$ with respect to D, $\triangle BCD$ with respect to A, and so on, as in Figure 16.2.

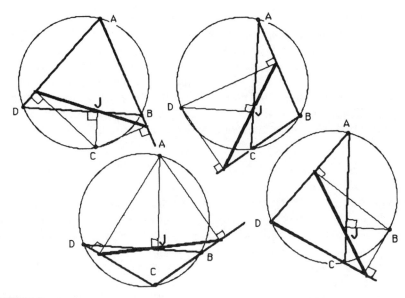

FIGURE 16.2.

After looking at some pictures like those in Figure 16.2, Dennis Johnson noticed that the four Simson lines seemed to be concurrent, and he proved that this was indeed the case. The result was known (Court, Theorem 304). Nevertheless, we will refer to this point of intersection as the *J* point of the given quadrilateral. No computers have been used so far, but Johnson next investigated the locus of the *J* point when △*ABC* is held fixed and the fourth point *P* moves around its circumcircle. A computer program produced pictures like those of Figure 16.3.

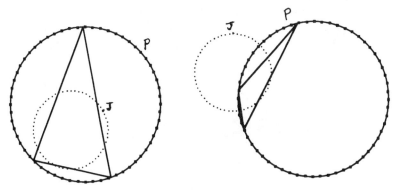

FIGURE 16.3.

As *P* runs around the original circle, the *J* point appears to run around the nine-point circle of the triangle. The nine-point circle is easy to identify: It is the circle through the midpoints of the three sides of the triangle (the other six points are the feet of the altitudes and the midpoints of the lines connecting the vertices to the orthocenter). Further, the *J* point moves around the nine-point circle in the same direction as *P* and with the same angular velocity. Johnson found proofs of all of these properties. Unfortunately, his write-up of the results was lost in the 1991 Oakland fire. The arguments, however, are not difficult if one knows some of the facts about the nine-point circle.

Next question: What does the family of all Simson lines for a given triangle look like as *P* runs around its circumcircle?

Johnson wrote a program that produced pictures of the "Simson family" for random triangles. Remarkably, all these line families looked the same, independent of the shape of the triangle. More precisely, all the line families appear to be congruent, differing only in their position and orientation with respect to the given triangle. The family always had an envelope that looked as if it might be a "hypocycloid of three cusps," that is, the locus of a point on a circle that rolls on the inside of a circle of three times its radius. Figure 16.4 presents two examples, one for an acute, the other for an obtuse, triangle. What is going on?

The answer once again involves the nine-point circle. When the graphics program was amended to include this circle, as shown in Figure 16.4, it was apparent that it was

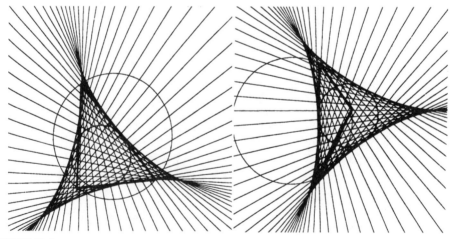

FIGURE 16.4.

the inscribed circle to the envelope of the Simson family. Observing this, Johnson worked out completely the choreography of the dance. The ballerina, Mlle. Simson, may be thought of as carrying a long balance bar, like those used by tightrope walkers. This represents the Simson line. At each instant, as P moves around this circle, the ballerina is at the J point and hence moves around the nine-point circle, say clockwise, with some uniform angular velocity ω. At the same time, she is required to rotate her body counterclockwise with angular velocity $\omega/2$, in a sort of slow-motion whirl. The balance bar then sweeps out the Simson family. This gives the entire story. The three cusps of the envelope occur when the balance bar goes through the center of the nine-point circle. We will call such a line a *cusp line*. Suppose initially the bar goes through the center of the circle, and let this cusp line be a reference line. Then when the radial line from the center to the J point has moved through an angle Θ, the balance bar will make an angle $(3/2)\Theta$ with the radial line. Hence it will pass through the center of the circle three times while the J point makes one revolution, producing the three cusp lines. The envelope turns out to be the hypocycloid obtained by rolling the nine-point circle (of radius $1/2$) on the inside of a circle of radius $3/2$.

The story is not quite finished. Note that the Simson hypocycloid is an invariant of the original triangle. It is natural to ask then, for example, how from the given triangle one finds, say, its cusp lines. A final surprise: these lines are, in general, not ruler-and-compass constructible. Johnson's nice argument shows that if the cusp lines *could* be constructed, then one could trisect an arbitrary angle, which, as we know, is not possible with ruler and compass.

CONFIGURATIONS WITH RATIONAL ANGLES

For background, the reader should read the intriguing article "Nineteen problems in elementary geometry" by Armando Machado in Appendix 3. In Figure 16.5, we reproduce the first of Machado's problems: Given the isosceles triangle with vertex angle 20° and lines a and b making angles of 60° and 50° with the base, determine the unknown angle γ. It is recommended that the reader take a moment to try to answer the question to appreciate its difficulties. It is not hard to write various trigonometric equations that γ must satisfy. Machado solved one of these numerically (using only a pocket calculator!) and was surprised to find that within the accuracy of the calculator, γ is 80°. Motivated by this discovery, he then did a numerical search using the software *Mathematica* and turned up several dozen other examples of these "rational configurations" where all of the angles are rational multiples of π. By lumping together certain families of solutions, Machado arrives at 19 distinct cases. Dennis Johnson made a computer search of a different sort which is described shortly, in which the lines a and b are allowed to meet the sides of the triangle not necessarily in the interior of the segments, and this turned up more than 250 examples of rational configurations. Some of these belong to various infinite linear families of configurations, whereas others seem to be isolated.

Of course, no numerical finding, no matter how convincing, constitutes a proof that these configurations actually exist. Machado's 19 problems are, therefore, to find the existence proofs. As is usual in geometry, there are two approaches, synthetic and algebraic.

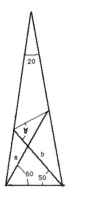

FIGURE 16.5.

The figure on the right in Figure 16.5 gives the picture for a synthetic proof for Machado's original example. Draw lines AB and BC, where AB makes an angle of 20° with the base. Then prove that all the marked segments are equal. From this and the known angles, the angle γ is obtained. In fact, Margarita Ramalho succeeded in finding synthetic proofs for the six Machado configurations with vertex angle 20°. The striking fact, however, was that each case required a different argument and a different set of as many as four auxiliary lines. It would, therefore, seem a hopeless task to try to find elementary synthetic proofs for all the examples that have been found, some, for example, whose angles $(k/n)\pi$ have denominators divisible by 7.

What one would like is a uniform procedure, an algorithm, for determining whether a rational configuration with given angles exists. As the unknown angle γ is uniquely determined by the given angles of Figure 16.5, it is clear that the four angles must satisfy some trigonometric equation. Johnson chose as parameters the angles shown in Figure 16.6. The relationship among the angles can be expressed by various equations. Johnson finds, for example,

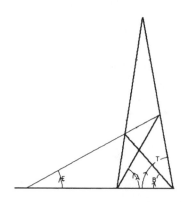

FIGURE 16.6.

$$(\sin^2 T) \sin (B + E - A) = \sin A \sin B \sin E.$$

It is useful to convert to complex exponentials. Letting $\alpha = e^{2iA}$, $\beta = e^{2iB}$, $\tau = e^{2iT}$, and $\epsilon = e^{2iE}$, one obtains

(1) $$(\tau - 1)(\tau^{-1} - 1)[(\alpha^{-1}\beta\epsilon) - 1] = (\beta - 1)(\epsilon - 1)(\alpha^{-1} - 1).$$

Now, if the numbers A, B, E, and T are rational multiples of π, then α, β, τ, and ϵ are roots of unity. For example, in the original problem, $A = (1/3)\pi$, $B = (5/18)\pi$, $T = (4/9)\pi$, and $E = (1/6)\pi$, with lowest common denominator 18. Taking ξ to be a primitive 18th root of unity, we get $\alpha = \xi^6$, $\beta = \xi^5$, $\tau = \xi^8$, and $\epsilon = \xi^3$. Substituting these in Eq. (1) gives the polynomial equation

(2) $$\xi^{17} + \xi^{15} - 2\xi^{12} + 2\xi^8 - \xi^5 - \xi^3 + \xi^2 - 1 = 0.$$

To show that the equation is satisfied, we check that it is divisible by $\xi^6 - \xi^3 + 1$, the cyclotomic polynomial for the primitive 18th roots of 1. In fact, Eq. (2) factors as

$$(\xi^6 - \xi^3 - 1)(\xi^{11} + \xi^9 + \xi^8 - \xi^6 - 2\xi^3 + \xi^2 - 1).$$

The method is clearly general and was used by Johnson to find all of his rational configurations. For each positive integer N, let A, B, T, and E take all integer values $< N$. Then substitute the corresponding powers of ξ in Eq. (1) and check to see whether it is divisible by the Nth cyclotomic polynomial. Unlike the numerical search, this method also *proves* that the configurations exist. Using a similar technique, Raphael Robinson also verified the existence of all of the configurations in Machado's article.

Despite the mass of data now available, the structure of the set of rational configurations remains quite mysterious. Johnson found rational configurations for all even values of N from 8 to 36, but there are none for odd N in this range (aside from such trivial cases as $\alpha = \beta$). Are there perhaps no nontrivial rational configurations with N odd, and if so, why? Also mysterious is the source of the original Machado example.

SOME LATE-BREAKING NEWS

I am grateful to Don Chakerian for providing me with some highly relevant information in connection with two of the items treated above. Regarding the section "The Dance of the Simson Lines" Chakerian writes,

> David Kay, *College Geometry*, Holt, 1969, mentions (p. 248) that the hypocycloid generated by those Simson lines was apparently first discovered by Jacob Steiner, Kay gives a development following E. H. Lockwood "Simson's Line and it's Envelope", *Math. Gaz.* 37 (1953), 124–125.

The second communication concerned the section "Configurations with Rational Angles". I concluded the above section by asking for information on the origin of the first problem in Machado's article, which he says "... must be rather well known as it appears repeatedly in mathematical circles." Thanks to Chakerian and also Stan Wagon, I was able to track down what appears to be the origin of the example, a problem in *Math. Gaz.* 11 (1922) proposed by E. M. Langley. More interesting is the fact that there is actually a small literature stemming from "Langley's problem." In fact all of the results in Machado's article and in the section above and a great deal more turns out to be already known. I will not attempt to give the complete bibliography (the best easy reference seems to be *Math. Gaz.* 62 (1978) 174–183), but the ultimate story is intriguing.

What might be called the generalized Langley problem is that of finding and classifying all "rational quadrangles," that is, all complete quadrangles such that the angle between any two of the six sides is a rational multiple of π. This problem was completely solved in 1978 by Paul Monsky, who showed that aside from the obvious examples, (e.g., a rational-angled triangle together with its angle bisectors), the solutions consist of 120 one-parameter families and 1,830 isolated cases. Monsky's manuscript, which ran some 30-odd typed pages, was never published because it turned out that his results had been anticipated 40 years earlier (!) by the Dutch geometer Gerrit Bol in a paper (in Dutch) "Beantwoording van Prijsvraag no 17," *Nieuw Arch. f. Wisk.* (2) 18 14–68 (1936).

By way of relating the Bol-Monsky results to some of those described above, it is not hard to show that the quadrangle problem is equivalent to asking when three or more diagonals of a regular n-gon are concurrent. This is illustrated for the original Langley problem in the figure below. It turns out that such concurrences cannot occur for n odd, and, except for obvious cases, can only occur for n divisible by 6. This confirms observations made by Dennis Johnson in his computer search, as reported above. Among other interesting results, Bol finds values of n for which 4, 5, 6, and 7 diagonals are concurrent and shows that these are the only possibilities.

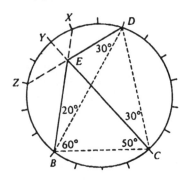

TRIANGLES WITHIN TRIANGLES

All right, boys and girls, today we're going to do an experiment in geometry. Now, I know you all know how to write computer programs that draw triangles and perpendicular lines, but today I'm going to show you a different method. All you need is a pencil (you remember them, don't you?) and a straightedge that can draw right angles. You each have a piece of paper showing the same triangle with sides a, b, and c. Now please follow my instructions. I want you to choose any point on your paper and draw a perpendicular to line a. Got it? From there, draw a perpendicular to line b, from there, a perpendicular to line c, and then a perpendicular back to line a, and keep on drawing perpendiculars, round and round. What do you notice? That's right, Karl Friedrich, after a while you keep getting the same triangle inside the one you started with. And, yes, it seems to be a smaller version of the original triangle turned on its side. (See Fig. 16.7.) Please notice that all of you got the same triangle even though you chose different starting points! Why is this, do you suppose? What's that, Henri? A "contraction mapping"? Hmmm.

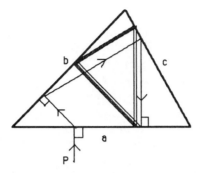

FIGURE 16.7.

All right, now we're going to try something a little different. Begin again with a new starting point, but this time instead of drawing your perpendiculars from *a* to *b* to *c*, go the other way, first to *a*, then to *c*, then to *b*, so you spiral around in the opposite direction. Right you are, Sonya, this time we get a different triangle, but it's really the one we got the first time, only it's standing on its head. (See Fig. 16.8.)

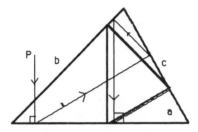

FIGURE 16.8.

These phenomena were discovered by Hidefumi Katsuura. The fact that the two limit triangles are similar to the original rotated through ±90° is clear, but Katsuura shows that they are, in fact, congruent. They have the same circumcircle and are related by symmetry with respect to the center of this circle. Further, the image of the original triangle under this symmetry contains the six vertices of the two limit triangles. All of this is illustrated in Figure 16.9.

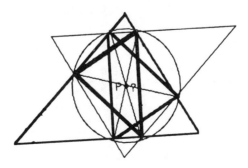

FIGURE 16.9.

Up to this point computers have not entered the picture. Note, however, that the center of symmetry P of Figure 16.9 is an invariant of the original triangle. The natural question arises, Which center is it? The centroid, the orthocenter, incenter, circumcenter, nine-point center? To answer this question, I did the obvious thing and dashed off an e-mail describing the construction to Clark Kimberling in Evansville, the previously mentioned world expert on centers of triangles. Kimberling ran a few numerical experiments and by return e-mail informed me than the point was none of the above, but that its coordinates agreed with those of the *symmedian,* or *Lemoine, point* to 14 decimal places!

What! You never heard of the symmedian point? The symmedian point is the *conjugate point* of the centroid. Conjugate point? OK, choose any point P that is not one of the vertices $A, B,$ and C of the triangle. Reflect lines $PA, PB,$ and PC in the angle bisectors at $A, B,$ and $C,$ respectively. The lines so obtained are (theorem) concurrent, and their intersection, $P',$ is the conjugate of $P.$

Knowing, or I should say suspecting, that P is the Lemoine point, one can prove analytically that it actually is, as did Clifford Gardner; or better still, one can turn again to N. A. Court, Chapter 10, "Recent Geometry of the Triangle," Section B, "Lemoine Geometry" (Emile Lemoine, 1840–1912), Theorem 593, from which the Katsuura results follow easily. The circle in the figure is known as *Lemoine's second circle,* and the three diameters are the *Lemoine antiparallels.* Katsuura, however, gives direct elementary proofs of his results, so knowledge of nineteenth-century mathematics is not needed.

ADDENDUM: JIGSAW PARADOXES

Our first section was devoted to demonstrating the power of the use of computer graphics in solving geometric problems. We conclude with a demonstration of the power of the *misuse* of computer graphics to provide fallacious solutions.

Figure 16.A1, which has been attributed to Lewis Carroll, is a well-known mathematical hoax. It claims to prove that area is not necessarily preserved under finite decomposition.

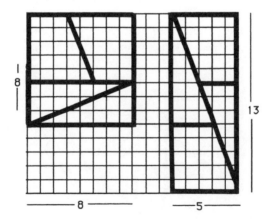

FIGURE 16.A1.

The reader who has not seen this before should try to find the fly in the ointment. The same general principle has been used to create other such paradoxical decompositions, such as that in Figure 16.A2.

A much more elaborate variation on this theme was invented recently by Jean Brette, who has been in charge of mathematics for many years at the *Palais de la Découverte,* the science museum of Paris.

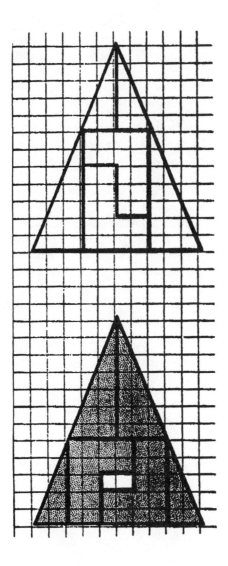

FIGURE 16.A2.

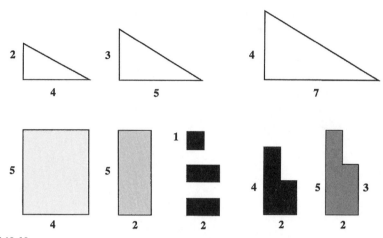

FIGURE 16.A3.

The example, illustrated in Figures 16.A3 and 16.A4, is actually six paradoxes in one. Using subsets of the 10 pieces shown in Figure 16.A3, one can assemble a 9 × 16 triangle in six different ways, corresponding to the six orderings of the three triangular pieces along the hypotenuse of the big triangle.

Referring to Figures 16.A3 and 16.A4, note that the nontriangular pieces of the bottom right triangle have total area of 44, those of the right-hand triangle have an area of 45, the next 46, the next 47, the next 48, and the lower left triangle 49. You might try making your own set of pieces out of cardboard. Amaze your friends!

In the dissection described here the trick is relatively apparent. However, Brette found a general method for constructing these paradoxical dissections and it is thus possible to construct examples where the three triangles along the hypotenuse are so similar to the big triangle that the discrepancy in shape becomes essentially imperceptible.

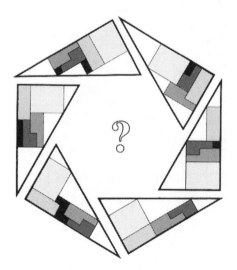

FIGURE 16.A4.

chapter 17

Packing Tripods

Sherman K. Stein

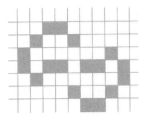

This chapter is devoted to an unsolved problem so simple to state that it can be told to the proverbial person in the street. Since it appears to be unrelated to known theorems, everyone, specialist or amateur, has an equal crack at it. The puzzle enthusiast, computer programmer, or mathematician may find it a tempting challenge. Moreover, because it hasn't been worked on by many people, there is a good chance for a fresh approach.

The problem, which concerns placing integers in the cells of a square array, grew out of a geometric question.

For a positive integer k, consider the tripod formed by a unit cube (the corner) to which arms of length k are attached at three nonopposing faces. The k-tripod consists of $3k + 1$ unit cubes. Figure 17.1 is a perspective view of a 4-tripod. The question is this: What fraction of the volume of space can be filled by nonoverlapping translates of a k-tripod when k is large?

Introduce an (x, y, z) coordinate system whose positive parts correspond to the directions of the arms of the tripods. The question leads to this related one: How many nonoverlapping k-tripods can have their corner cubes in the cube of side k, $0 \leq x, y, z \leq k$?

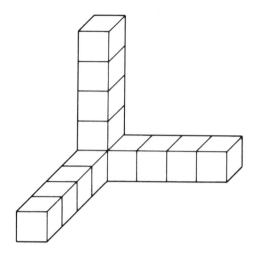

FIGURE 17.1.

Call this number $f(k)$. It is known that if $f(k)/k^2$ approaches 0 as k increases, then the fraction of space that can be packed by the k-tripods also approaches 0. So we now have the question, Does $f(k)/k^2$ approach 0 as k increases?

It is not hard to show that we may assume that each corner cube coincides with one of the k^3 unit cubes that make up the k-cube. Identify each of these unit cubes with its vertex that has the largest coordinates, hence with a triplet of integers (x, y, z), $1 \leq x, y, z \leq k$. Because the tripods don't overlap, for a given pair (x, y), there is at most one number z such that the triplet (x, y, z) is present. Therefore, we can record the presence of a tripod in a packing by entering the number z in the unit cell corresponding to the coordinates (x, y). Because the tripods don't overlap, the resulting entries satisfy the three conditions in the following definition of a "monotonic matrix":

Let k be a positive integer. An *array of order* k consists of k^2 empty cells arranged in a $k \times k$ square. Place in some of these cells any one of the numbers $1, 2, \ldots, k$, subject to three rules:

1. In each (vertical) column the entries strictly increase in size from bottom to top.
2. In each (horizontal) row the entries strictly increase from left to right.
3. The cells occupied by any specific integer rise as we move from left to right (the "positive slope" condition).

Call such an array, with some cells filled according to the three rules, a *monotonic matrix of order* k. From now on we deal with monotonic matrices instead of packings by tripods.

Then $f(k)$ is the maximum number of cells occupied in any monotonic matrix of order k. Clearly, $f(k) \leq k^2$. Moreover, since you could place the numbers $1, 2, \ldots, k$ in order in a single row, $f(k) \geq k$. The question is, What happens to the quotient $f(k)/k^2$ as k gets arbitrarily large? Does it have a limit? Is this limit zero?

To get a feel for $f(k)$, consider a few small values of k. Figure 17.2 illustrates the cases $1 \leq k \leq 5$. Clearly $f(1) = 1$, and $f(2) = 2$. It takes a little time to show that $f(3) = 5$ and that $f(4) = 8$. The case $k = 5$ was settled by an exhaustive computer search programmed

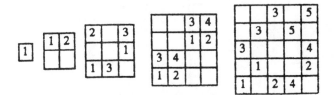

FIGURE 17.2.

by K. Joy, which showed that $f(5) = 11$. One of the many solutions found is displayed in Figure 17.2. These are the only values of k for which $f(k)$ is known.

When k is a square, there is always a monotonic matrix of order k with $k^{3/2}$ occupied cells. Figure 17.3 and the fourth array in Figure 17.2 illustrate the construction. This suggests writing $f(k)$ in the form $f(k) = k^{e(k)}$ and studying the behavior of $e(k)$. At first it was conjectured that $e(k)$ approaches $3/2$ as k increases. [This would imply that $f(k)/k^2$ approaches 0.] It is known that $e(k)$ does approach some number. Clearly, this number is no larger than 2. Call it L. It turns out that L is the smallest number that is greater than or equal to all the $e(k)$'s.

						7	8	9
						4	5	6
						1	2	3
			7	8	9			
			4	5	6			
			1	2	3			
7	8	9						
4	5	6						
1	2	3						

FIGURE 17.3.

With this in mind, consider Figure 17.4. This monotonic matrix shows that $f(7) \geq 19$. Taking logarithms of both sides of the equation shows that $e(7) \geq (\log 19)/(\log 7) \approx 1.513$. Thus L is at least 1.513.

If you examine Figure 17.4, you will note that it is based on the monotonic matrix of order 3 shown in Figure 17.2, with each of the nine blocked-out areas in Figure 17.4 corresponding to a cell in an array of order 3. The same technique provides the monotonic matrix of order 9 with 28 occupied cells (one more than in Fig. 17.3) shown in Figure 17.5. This implies that $e(9) \geq 1.516$, hence that $L \geq 1.516$. The idea behind the construction of Figures 17.4 and 17.5 shows that

$$f(2k + 1) \geq 2f(k) + 3k.$$

		4		6		7
	4					5
4				5	7	
				1	2	3
2		3	7			
		1	6			
1	3		5			

FIGURE 17.4.

			5				8	9
		5					6	7
	5				8	9		
5					6	7		
					1	2	3	4
		3	4	9				
		1	2	8				
3	4			7				
1	2			6				

FIGURE 17.5.

Let k and l be positive integers. By cutting an array of order kl into l^2 blocks of size $k \times k$, you can show that

$$f(kl) \geq f(k)\, f(l).$$

It is also clear that $f(k+1) \geq f(k) + 1$. It is these last two inequalities, together with the fact that $f(k) \leq k^2$, that imply by a known theorem that $e(k)$ has a limit as k increases.

There are several possible next steps in determining the behavior of $f(k)$. One is to use a computer to determine some more values of $f(k)$ or at least to find larger values of $e(k)$. D. R. Hickerson has shown (without computer) that $f(255)$ is at least 4638, which implies that $L \geq 1.523$. However, there may be a monotonic matrix of a small order that implies that L is even larger.

Until someone discovers what happens to $f(k)/k^2$ as k increases, following an honored tradition, I will propose what may be a simpler "practice" problem, also unsolved.

Let j be a fixed positive integer. Let $g(j, k)$ be the maximum number of cells in an array of order k that can be occupied by the numbers $1, 2, \ldots, j$, subject to the three rules given earlier. It is known that for each j, $g(j, k)/k$ approaches a limit as k increases, which will be denoted by $c(j)$.

First of all, $g(1, k) = k$. (Put 1s on the upward sloping diagonal of an array of order k.) Thus $c(1) = 1$. For $j = 2$, insert 1s and 2s as indicated in Figure 17.6. This construction, which provides four entries per three columns, shows that $c(2)$ is at least 4/3. It is known that $c(2) = 4/3$. It is also known that $c(3) = 5/3$ and that $c(4) = 2$. Thus, for $j = 1, 2, 3$, and 4, $c(j) = (k + 2)/3$. This pattern suggests that $c(5)$ should be 7/3, but all that is known is that it lies somewhere between 16/7 and 5/2.

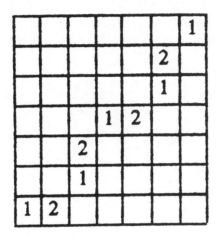

FIGURE 17.6.

By the way, one could define an n-pod in n-dimensional space. It consists of a corner cube with n arms of length k glued at nonopposite faces. (When n is 2, it looks like the letter L.) In dimensions 1 and 2 it tiles the space. As D. R. Hickerson pointed out, if $f(k)/k^2$ approaches 0, then in all dimensions from 3 on, n-pods pack n-space with a density approaching 0 for large k. So 3 is the critical dimension. Of course, I don't know if $f(k)/k^2$ approaches 0 as k increases. In fact, I don't even know that it has a limit.

REFERENCES

1. W. Hamaker and S. Stein, *IEEE Trans. Inform. Theory* 30, 364–368 (1984).
2. S. Stein and S. Szabo, *Algebra and Tiling,* Mathematical Association of America, Washington, D.C., 1994, Chapter 3.

chapter 18

Further Travels
with My Ant

David Gale, Jim Propp,
Scott Sutherland,
and Serge Troubetzkoy

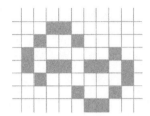

INTRODUCTION

A recurring theme of this book has been computer-generated mysteries. Examples are sequences defined by simple rational recursions whose terms turn out to be integers with interesting but unexplained divisibility properties or geometric configurations that exist although there are no proofs of existence. In most of the examples, the reported mysteries have remained unsolved and in some cases may in fact be, in a suitable sense, unsolvable. It is therefore gratifying to be able to present an elegant solution of a previously described mystery. An especially pleasing feature of this solution is that the breakthrough became possible by drawing the right picture. Once the picture is drawn, it becomes clear what must be proved, after which further study of the picture gives the clue for constructing the proof. It turns out that at one point one needs to use the Jordan curve theorem for a special class of closed curves.

In the following paragraphs we present an informal but (we hope) convincing argument that requires little more of the reader than carefully observing a collection of pictures.

The Story So Far and the Mystery

We are concerned once again with a certain automaton called an ant. An ant lives in a plane partitioned in the standard way into squares, or, as we call them, *cells* (think of the corners of the cells as the lattice points of the plane). Each cell is in one of several states, and the respective states of these cells change over time as the cells are visited in turn by the ant. Before we explain precisely how the ant interacts with its surroundings, it will help to consider a specific example. In the simplest nontrivial version of the ant, originally studied by Chris Langton, there are just two states, which we will call L (for left) and R (for right). The ant starts on the boundary between two cells, heading in one of the four cardinal compass directions (east or west on the vertical boundaries and north or south on the horizontal boundaries). As the ant advances through the cell, it makes a 90° turn, turning left in L-cells and right in R-cells, and moves to the boundary of the neighboring cell. As the ant leaves each cell, it changes the state of the cell, switching L-cells to R-cells, and vice versa. For an informal discussion of this ant see [1], where a number of interesting ant behaviors are observed. For our present purposes we note only the following still unexplained phenomenon: If initially all cells were in the same state, then at various times the ant's "track" (the set of cells it has visited, together with their current states) is centrally symmetric, that is, the configuration of L- and R-cells has central symmetry. In Figure 18.1 we reproduce the corresponding pictures, where the R-cells are black and the L-cells white. All cells were initially in the L-state, and the ant started out heading west on the right boundary of the central cell and about to enter it. These central symmetries stop occurring eventually, and after about 10,000 time-units, the ant settles into a periodic "highway-building" behavior, heading off to infinity in a southwesterly direction. This phenomenon of "transient symmetry" awaits a satisfying explanation.

In [2], Jim Propp describes the behavior of the more general n-state ants. There are now n different states for cells, numbered 1 through n, and it is the ant's internal "programming" that tells it which states to treat as L's and which states to treat as R's in choosing which way to turn. We represent this programming by means of a *rule-string*, a string of n L's and R's whose kth letter represents the ant's course of action when it comes to a cell in state k. For instance, the seven-state string LLRRRLR represents an ant that turns

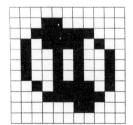 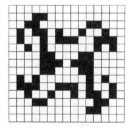 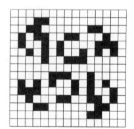

FIGURE 18.1. The universe of ant 2 at times 184, 368, and 472.

left on visiting cells in states 1, 2, and 6 and turns right on visiting cells in states, 3, 4, 5, and 7. When an ant comes to an L-cell (a cell in an L-state), it turns left. When it comes to an R-cell, it turns right. When the ant leaves a cell that is in state i, the cell changes to state $i + 1$ (mod n). In this terminology, the simple ant has the rule-string LR.

From considerations of symmetry it clearly suffices to consider only strings that start with an L. If we replace an L by a 1 and an R by 0 in the rule-string, we see that each ant corresponds to a positive integer expressed in base 2. So the simple ant is ant 2, and our seven-state ant with "genome" LLRRRLR is ant 98. Propp finds that different ants behave very differently depending on their rule-string. Some seem to be completely chaotic, whereas others eventually build highways. The only general statement that can be made is the following:

Fundamental Theorem of Myrmecology (Bunimovich–Troubtezkoy). *An ant's track is always unbounded, provided that its rule-string contains at least one L and one R.*

(For a proof see [1].)

In this exposition we concentrate on the phenomenon described by Propp as follows:

> Ants 9 and 12 [LRRL and LLRR in our notation] are the truly surprising ones [among ants with rule-strings of length 4]. In each case the patterns get ever larger, but without ever getting too far away from bilateral symmetry! More specifically, one finds that the ant makes frequent visits to the cell it started from and when it does, the total configuration quite often has bilateral symmetry.

Figures 18.2 and 18.3, reproduced from [2], show the track of ant 12 after 16,464 steps and the track of ant 9 after 36,836 steps. The figures use black, white, and two shades

FIGURE 18.2. The universe of ant 12 at time 16,464.

FIGURE 18.3. The universe of ant 9 at time 38,836.

of gray to indicate the four different states. Propp continues, "To find more ants of this sort we have to move to rule-strings of length 6. Here we encounter another mystery: the rule-strings that lead to bilaterally symmetric patterns are 33, 39, 48, 51, and 60. Note that all these numbers are divisible by three! Surely this cannot be an accident." Indeed it is not, as we will presently show.

TRUCHET TILES

First a preliminary observation. Since an ant turns through a right angle after each move, it follows that its moves are alternately horizontal and vertical. Therefore, the cells of the plane split up in checkerboard fashion into two sets: H-cells, which the ant always enters horizontally, from the left or the right, and exits vertically, either at the top or the bottom; and V-cells, which the ant always enters vertically and exits horizontally. Now, a crucial property of a cell is that it can change its state and thus change the course of action (left turn versus right turn) that the ant will take after its next visit to the cell. The key idea for representing this property, due to Bernd Rümmler, was to actually install schematic "switches" in the cells, where the orientation of the switch indicates whether the cell is

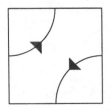
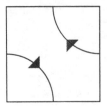

FIGURE 18.4. H-cells in L and R orientation.

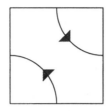 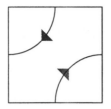

FIGURE 18.5. V-cells in L and R orientation.

currently in an L-state or an R-state. These switches, called Truchet tiles, are illustrated in Figure 18.4, which shows an H-tile first in its L-orientation and then in its R-orientation, and Figure 18.5, which shows the same thing for a V-tile (An "H-tile" is the Truchet tile associated with an H-cell, and similarly for V-tiles.) Note that switching orientation corresponds to reflecting the tile about either its vertical or horizontal axis.

Figure 18.6 shows the plane paved with Truchet tiles, all of which have the L-orientation, giving a pattern of disjoint circles: the "initial configuration." As the ant moves, some of the tiles switch in accordance with the given rule-string, but it is clear that the pattern will always consist of a set of disjoint simple closed curves, which we call *Truchet contours*. (Infinite contours cannot arise, since at each stage the ant has switched only finitely many of the tiles.) It will be helpful (at least for now) to imagine that the ant actually travels along the Truchet curves themselves, rather than along a lattice path joining the centers of the cells, and that the ant's initial position is the midpoint of one of the edges of the central Truchet tile. Figure 18.7 gives the "Truchet pictures" corresponding

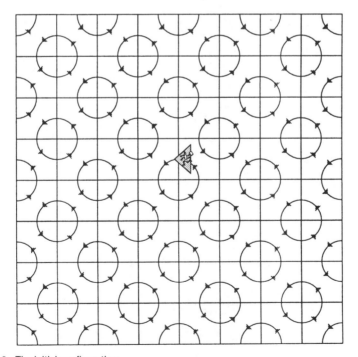

FIGURE 18.6. The initial configuration.

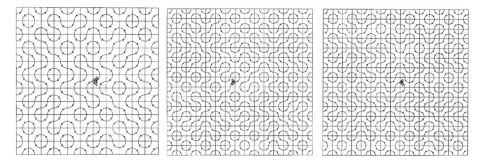

FIGURE 18.7. The universe of ant 2 at times 184 (top left), 368 (left), and 472 (right).

to the (transiently) centrally symmetric configurations of the simple ant given by Figure 18.1, whereas Figures 18.8 and 18.9 give Truchet pictures corresponding to Figures 18.2 and 18.3, and manifest the same bilateral symmetry. In all of these pictures the ant has returned to its original location, and of particular interest is the Truchet contour through this point, which we call the *principal contour*. Initially, the principal contour is just a circle, as are all other contours. In Figure 18.9 the principal contour has been highlighted.

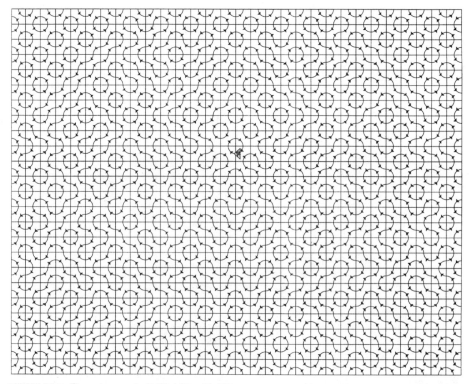

FIGURE 18.8. The universe of ant 12 at time 16,464.

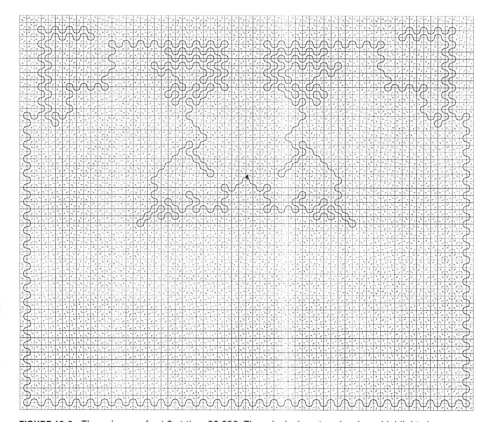

FIGURE 18.9. The universe of ant 9 at time 38,836. The principal contour has been highlighted.

Now, if one knew that each time the ant left its starting location it would stay on the principal contour until it returned once more to its starting location, then it would follow that when the ant had completed its tour, all the symmetries of the initial state of the universe would be preserved, for whenever a cell was visited, its symmetric mate would also be visited.

Unfortunately, an ant will not, in general, stay on the principal contour, because (as in Fig. 18.10) the contour may pass through some cells twice. This means that when the

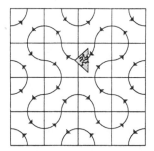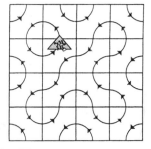

FIGURE 18.10. Switching: the universe of ant 2 at times 4 and 5.

ant returns to such a cell, the orientation of that cell may have switched (indeed this will always be the case for the simple ant studied by Langton), causing the ant to leave the principal contour. However, for ants listed by Propp, one can show that, in fact,

(1)   *The cells that are visited twice by a contour never switch on thr first visit.*

This implies that the ant engages in a process of repeatedly tracing out bilaterally symmetric principal contours, resulting in a bilaterally symmetric universe each time the ant returns to its starting point. Property (1) was first proved by Rümmler for ant 12 and then generalized to the other ants by Troubetzkoy. The argument to follow is a reworking of those proofs by Propp.

THE EVEN RUN-LENGTH PROPERTY AND THE AUGMENTED PICTURE

Why do some ant tracks exhibit recurrent bilateral symmetry and others not? Here are the rule-strings for the 4- and 6-state ants of Propp's article that exhibit recurrent symmetry:

Ant	Rule-String
9	LRRL
12	LLRR
33	LRRRRL
39	LRRLLL
48	LLRRRR

Ant	Rule-String
51	LLRRLL
57	LLLRRL
60	LLLLRR

It is not hard to see what these strings have in common. If we think of them as arranged in cyclic rather than linear order, then each of them consists of an even number of L's followed by an even number of R's. In general, we say that a rule-string has the *even run-length property* if in the cyclic order it consist of alternate runs of L's and R's of even length, for example, LRRLLRRRRL. For simplicity in what follows, we will consider only the case where the string starts with an even number of L's (ants 12, 48, 51, and 60). The argument for the other case is similar.

Why does the even run-length property imply recurrence of bilateral symmetry? Here Rümmler has augmented the picture in a manner paraphrased as follows. Let us say that a cell is *cold* if its state is odd, so that it will not change orientation the next time it is visited (because of the even run-length property) and *hot* if its state is even, so that it may or may not change orientation the next time it is visited. To complete the picture we adopt the convention that for hot cells we display not only the Truchet tile but also its diagonal. The *diagonals graph* is the graph whose edges consist of the diagonals of the hot tiles. Figures 18.11 and 18.12 show the diagonals graphs of ants 12 and 48 associated with certain

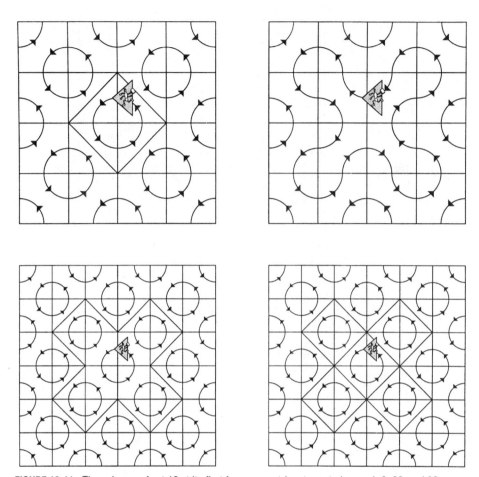

FIGURE 18.11. The universe of ant 12 at its first four symmetric returns to home: 4, 8, 28, and 32.

instants at which the ant has returned to its initial location (hereafter called "home"). These graphs can be quite complicated, breaking up into many components, as shown in Figure 18.12. However, observe a key fact, which we call the *even diagonals-degree property*:

(2) All vertices in the diagonals graph have even degree (thus, degree 0, 2, or 4).

Lemma 1. *Suppose that the ant is at its home position and that the state of the universe satisfies* (2). *Then the ant will travel along the principal contour (and return home).*

Proof: As remarked earlier, the only thing we have to worry about is that the principal contour *C* might visit a cell twice and that this cell might change its orientation after the ant's first visit. If the twice-visited cell is cold, then the Truchet tile does not change its orientation after the ant's first visit. What about twice-visited cells that are hot? We will show that such cells do not exist, as a consequence of (2). Let *T* be such a cell, and let *d* be the diagonal of *T* connecting vertices *u* and *v*.

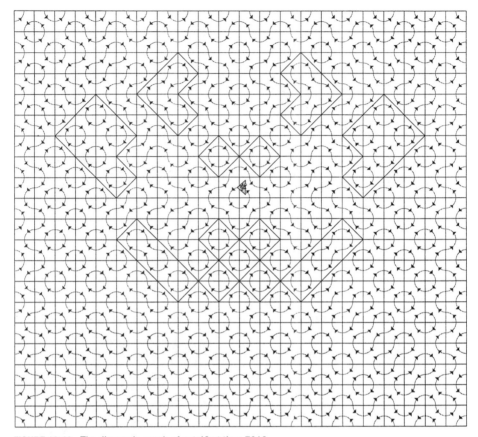

FIGURE 18.12. The diagonals graph of ant 48 at time 7016.

Claim. *If d is deleted, then in the resulting graph, the components of u and v are disjoint.*

If we can show this, then the desired contradiction follows, since from (2) the component of *u* (or *v*) would contain only one vertex of odd degree, contradicting the well-known fact (usually associated with Euler) that a connected graph must always have an even number of vertices of odd degree. (This is sometimes referred to as the handshake theorem, which says that the number of people who have shaken hands an odd number of times is even.)

It remains to prove the claim. For this purpose consider the twice-visited tile *T* and its two arcs (quarter circles). Without loss of generality, we assume that *T* is an H-tile. Now color in red the arc in *T* below the diagonal and, in addition, all succeeding arcs of *C* up to the point where *C* is about to reenter *T*. Color the remaining arcs blue. The dotted arcs in Figures 18.13(a) and 18.14(a) represent the red arcs, and the solid arcs represent the blue ones. Now consider the same picture, except that the diagonal of *T* has been deleted and *T* has switched to its other orientation, as shown in Figures 18.13(b) and 18.14(b). As before, the arc in *T* below the diagonal should be colored red and the other arc blue. One sees at once that *C* has split into two contours: one all red, the other all blue (this much is

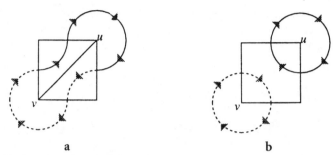

FIGURE 18.13. The nonnested case: (a) before; (b) after.

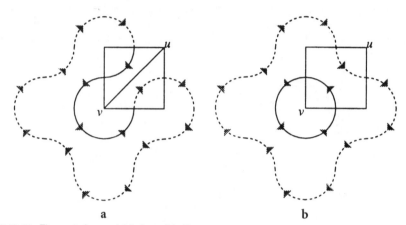

FIGURE 18.14. The nested case: (a) before; (b) after.

a purely combinatorial fact and has nothing to do with the topology of the plane). Now the Jordan curve theorem tells us that each of these nonintersecting contours has an inside and an outside, which can be arranged either as in Figure 18.13(b) (the nonnested case) or as in Figure 18.14(b) (the nested case). In either case it is clear that the components of u and v are disjoint, since both the red contour and blue contour intervene. Combining this with the handshake theorem completes the proof.

Remark. Although these last observations can be made rigorous without using the full force of the Jordan curve theorem, there must be some use of the topology of the plane, for the analogue of Lemma 1 is not true on the torus.

Finally, we must prove (2).

Lemma 2. *If (2) holds when the ant is at home, it will still hold after the ant has toured the principal contour and returned home.*

Proof. Let v be some vertex of the diagonals graph. The *neighborhood* $N(v)$ consists of the four cells having v as a vertex. We show that if at some point the Truchet contour enters and then leaves $N(v)$, the parity of the degree of v is not changed.

Case I. The contour meets $N(v)$ in only one cell. Then v does not lie on a diagonal of that cell. If the cell is cold, the situation remains the same, since cold cells do not switch. If it is hot, then it becomes cold, and hence has no solid diagonal. So, whether it switches or not, the degree of v is unchanged.

Case II. The contour meets more than one cell of $N(v)$. Then let E be the cell where it enters and E' the cell where it exits. If E is hot, then its diagonal is incident on v. So when it becomes cold, this diagonal disappears (whether E switches or not). If it is cold, it becomes hot (without switching), so a new solid diagonal will be incident on v. The exact same argument applies to E'. So the net effect of the two changes is to preserve the parity of the degree of v (see Fig. 18.15). As for intermediate cells (there may be one or two of them), their diagonals do not meet v, so the argument is the same as for Case I. These cells do not contribute to any change in the number of solid diagonals incident on v. There is an additional case in which the cells E and E' coincide but we leave the analysis to the reader.

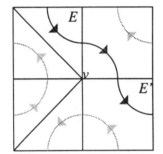 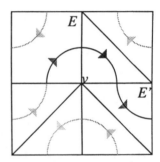

FIGURE 18.15. The parity of the number of solid diagonal incident to v does not change.

In general, the principal contour may reenter and reexit $N(v)$ several times, but if one looks at the portion of the curve between any entrance point and the corresponding exit point, the preceding analysis applies. This completes the proof.

Combining Lemmas 1 and 2, we finally see what is happening: If the state of the universe satisfies the even diagonals-degree property with the ant at home, then the ant must travel along the principal contour, but when it completes this path and returns home, it restores the even diagonals-degree property, so that it must once again travel along the (new) principal contour, and so on, *ad infinitum*. In view of the argument at the end of the preceding section, the proof of the theorem (and the solution of the symmetry mystery) is complete.

We leave it to the reader to find the simple number-theoretic argument that shows that a number like 57 whose binary expansion has the even run-length property is divisible by 3. This explains Propp's observation concerning the code numbers of the ants whose tracks exhibit recurrent bilateral symmetry.

Note: *For a copy of Propp's program ant.c (an ant-universe simulator designed for UNIX machines), send e-mail to propp@math.mit.edu.*

REFERENCES

1. D. Gale. "The industrious ant," *Mathematical Intelligencer* 15(2), 54–58 (1993) (Chapter 10 of this book).
2. D. Gale and J. Propp, "Further ant-ics," *Mathematical Intelligencer* 16(1), 37–42 (1994) (Chapter 13 of this book).
3. L. A. Bunimovich and S. Troubetzkoy, "Recurrence properties of Lorentz Lattice Gas Cellular Automata," *J. Statist. Phys.* 67 289–302 (1992).

chapter 19

The Shoelace Problem

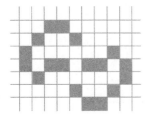

Has this ever happened to you? You've just finished patiently trying to explain some beautiful result of pure mathematics to a group of non-mathematicians, hoping that you've conveyed something of the flavor of this pearl of truth and beauty, and then after a pause someone says, "Yes, but what has any of this got to do with everyday life?" After much thought I've decided that the correct response is to say "Nothing. That's what's so nice about it. After all, everyday life is often a drag, so we do mathematics for the same reason we listen to music or ski down a mountain: to get away from and above and beyond everyday life."

But now I must admit that every once in a while it works out the other way and EDL turns out to be a source of unexpectedly interesting mathematics. A nice example of this is this chapter written by John H. Halton.

John H. Halton

In a number of discussions about how shoes should be laced, it became apparent that no one seemed to have the definitive answer. Shoes were laced and relaced, passions flared and shoes were even thrown. … The author decided that an appeal to mathematics was indicated.

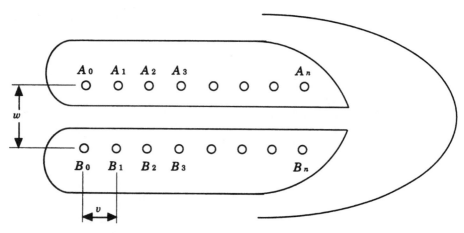

FIGURE 19.1. The shoe (a schematic).

This problem is a restriction of the traveling salesman problem. We are given a set of $2(n+1)$ points (the *laceholes*, or *eyelets*) arranged in a bipartite lattice, as shown in Figure 19.1.

The problem is to find the shortest path from A_0 to B_0, passing through every eyelet just once, so that points of the subsets

(1)
$$A = \{A_0, A_1, A_2, \ldots, A_n\} \quad \text{and}$$
$$B = \{B_0, B_1, B_2, \ldots, B_n\}$$

alternate in the path.

The three standard lacing strategies are shown in Figures 19.2–19.4.

For the American (AM) style, as in Figure 19.2, if n is *odd*, the lacing is

(2)
$$A_0 \to B_1 \to A_2 \to B_3 \to A_4 \to \cdots$$
$$\to A_{n-1} \to B_n \to A_n \to B_{n-1} \to A_{n-2} \to B_{n-1} \to \cdots$$
$$\to A_3 \to B_2 \to A_1 \to B_0.$$

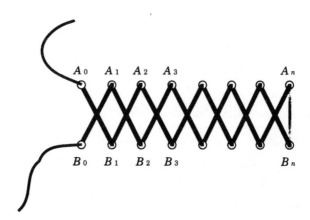

FIGURE 19.2. American zigzag lacing.

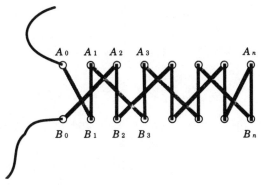

FIGURE 19.3. European straight lacing.

If n is *even*, the lacing, similarly, is

$$
\begin{aligned}
&A_0 \to B_1 \to A_2 \to B_3 \to A_4 \to \cdots \\
&\to A_{n-2} \to B_{n-1} \to A_n \to B_n \to A_{n-1} \to B_{n-2} \to \cdots \\
&\to A_3 \to B_2 \to A_1 \to B_0,
\end{aligned}
$$

(3)

and it is easily verified that in either case, the total length of lace used is given by

(4)
$$
L_{AM} = L_{AM}(n, v, w) = w + 2n\sqrt{v^2 + w^2}.
$$

For the European (EU) style, as in Figure 19.3, when n is *odd*, the lacing is

(5)
$$
\begin{aligned}
&A_0 \to B_1 \to A_1 \to B_3 \to A_3 \to \cdots \\
&\to A_{n-2} \to B_n \to A_n \to B_{n-1} \to A_{n-1} \to B_{n-3} \to \cdots \\
&\to B_2 \to A_2 \to B_0.
\end{aligned}
$$

When n is *even*, the lacing, similarly, is

(6)
$$
\begin{aligned}
&A_0 \to B_1 \to A_1 \to B_3 \to A_3 \to \cdots \\
&\to A_{n-1} \to B_n \to A_n \to B_{n-2} \to A_{n-2} \to B_{n-4} \to \cdots \\
&\to B_2 \to A_2 \to B_0,
\end{aligned}
$$

and with a little more thought, we see that in both cases, the total length of lace is given by

(7)
$$
L_{EU} = L_{EU}(n, v, w) = nw + 2\sqrt{v^2 + w^2} + (n-1)\sqrt{4v^2 + w^2}.
$$

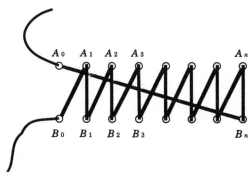

FIGURE 19.4. Shoe shop quick lacing

For the shoe shop (SS) style, as in Figure 19.4, the lacing is

(8)
$$A_0 \to B_n \to A_n \to B_{n-1} \to A_{n-1} \to \cdots$$
$$\to B_3 \to A_3 \to B_2 \to A_2 \to B_1 \to A_1 \to B_0,$$

and we find that the total length is given by

(9)
$$L_{SS} = L_{SS}(n, v, w) = nw + n\sqrt{v^2 + w^2} + \sqrt{n^2 v^2 + w^2}.$$

We can generalize the situation as follows. Let α and β denote permutations of $\{1, 2, 3, \ldots, n\}$:

(10)
$$\alpha = \{\alpha_1, \alpha_2, \ldots, \alpha_n\},$$
$$\beta = \{\beta_1, \beta_2, \ldots, \beta_n\},$$

To these will correspond the lacing

(11)
$$A_0 \to B_{\beta 1} \to A_{\alpha 1} \to B_{\beta 2} \to A_{\alpha 2} \to B_{\beta 3} \to \cdots$$
$$\to A_{\alpha n-1} \to B_{\beta n} \to A_{\alpha n} \to B_0,$$

which has total length

(12)
$$L = \sqrt{\beta_1^2 v^2 + w^2} + \sqrt{(\alpha_1 - \beta_1)^2 v^2 + w^2}$$
$$+ \sqrt{(\beta_2 - \alpha_1)^2 v^2 + w^2} + \sqrt{(\alpha_2 - \beta_2)^2 v^2 + w^2}$$
$$+ \cdots + \sqrt{(\beta_n - \alpha_{n-1})^2 v^2 + w^2} + \sqrt{\alpha_n^2 v^2 + w^2}.$$

For the three special lacings shown above, the particular permutations are

(13)
$\alpha_{AM} = \{$all even numbers increasing; then all odd numbers decreasing$\}$,
$\beta_{AM} = \{$all odd numbers increasing; then all even numbers decreasing$\}$;

(14) $\alpha_{EU} = \beta_{EU} = \beta_{AM}$,

(15) $\alpha_{SS} = \beta_{SS} = \{$all numbers decreasing.$\}$

The simplicity of these permutations is indeed remarkable.

Theorem 1. *If $v = 0$ or $w = 0$, then for all positive n,*

(16)
$$L_{AM} = L_{EU} = L_{SS}.$$

If $v \geq 0$ and $w \geq 0$,

(17)
$$L_{AM}(1, v, w) = L_{EU}(1, v, w) = L_{SS}(1, v, w),$$

and if $v > 0$ and $w > 0$, then

(18)
$$L_{AM}(2, v, w) < L_{EU}(2, v, w) = L_{SS}(2, v, w),$$

Finally, if $v > 0$, $w > 0$, and $n > 2$, then

(16)
$$L_{AM} < L_{EU} < L_{SS}.$$

This theorem can be proved, using (4), (7), and (9), by carefully analyzing cases and eliminating radicals. The proof is left as an exercise for the reader. (It is given by the author in a technical report [1].)

THE LATTICE REPRESENTATION

Let us make a lattice of alternating parallel, equidistant sets A and B, as shown in Figure 19.5. Given any lacing $\mathcal{L}$, we can represent it, as is shown for our three standard examples, by a polygonal (piecewise straight) line L moving always downward across the new lattice, visiting the eyelets only once each.

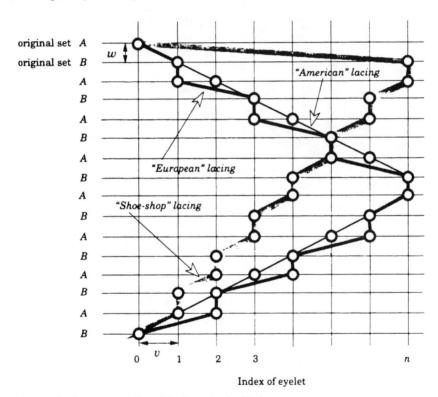

FIGURE 19.5. Lattice representation of the three standard lacings.

The first line segment in the order of lacing, $A_0 \to B_{\beta 1}$, is unchanged; the next, $B_{\beta 1} \to A_{\alpha 1}$, is replaced by its mirror image in the original B line; the next, $A_{\alpha 1} \to B_{\beta 2}$, is moved downward by two lattice intervals, parallel to itself (i.e., it is a twice-repeated mirror image), and so on; the last segment, $A_{\alpha n} \to B_0$, returns to the image of B_0 in the B line displaced downward by $2n$ intervals. Clearly, the total length of the representation L equals the original total length of L, the lacing $\mathcal{L}$ itself.

That the "American" (AM) lacing is shorter than the "European" (EU) lacing is now immediately apparent by a straightforward application of the triangle inequality (see Figure 19.6).

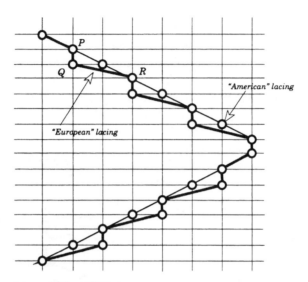

FIGURE 19.6. Comparison of AM and EU lacing.

The two representations L_{AM} and L_{EU} coincide in several places. Where they differ, replicas of a triangle PQR occur, and it is clear that $PR < PQ + QR$, so that the first inequality in (19) follows, without further algebra!

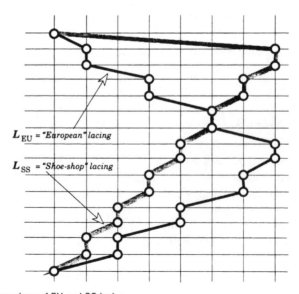

FIGURE 19.7. Comparison of EU and SS lacing.

That the EU lacing is better than the SS lacing is a little harder to prove (see Figure 19.7). First, we observe that the representations L_{EU} and L_{SS} have just *two* diagonal segments in common, moving by one lattice interval in both directions (slopes $\pm w/v$), and n (vertical) segments, moving by one vertical lattice interval w only. If we omit all of these common intervals, shifting the separated lower segment upward (and in the first two cases, sideways also), parallel to themselves, to rejoin the upper segment, and thus *subtracting equal lengths from each representation,* we obtain reduced representations L_{EU}^* and L_{SS}^*. The result is shown in Figure 19.8. Each representation now consists of a singly broken line (just two successive line segments—a *zig* and a *zag*).

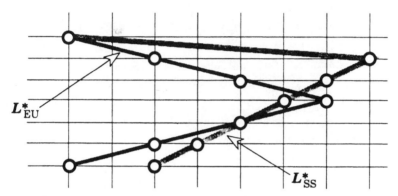

FIGURE 19.8. Comparison of EU and SS lacing—reduced representations.

Now perform the "reflection trick" again, this time in the horizontal coordinate direction, so that the leftward segment of each representation is reflected about the vertical. The resulting representation lines are denoted by L_{EU}^{**} and L_{SS}^{**} (see Fig. 19.9).

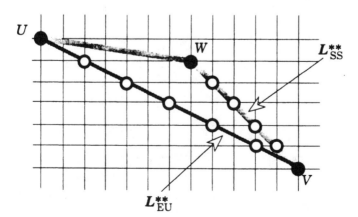

FIGURE 19.9. Comparison of EU and SS lacing—reflected representations.

We can now simply observe that L_{EU}^{**} is just a single straight segment UV, whereas L_{SS}^{**} consists of two straight segments, UW and WV, so that again by the triangle inequality, (19) clearly holds.

Optimization

We adopt the lattice representation described above (see Figures 19.5–19.7) and apply the "reflection trick" to the part of the path from B_n to B_0. The form of the path corresponding to a typical general lacing is illustrated in Figure 19.10. The path L_{AM} corresponding to the AM lacing is also shown. In this particular example, as before, $n = 7$ and the lacing is

(20)
$$A_0 \to B_2 \to A_7 \to B_4 \to A_6 \to B_1 \to A_1 \to B_3 \to A_3$$
$$\to B_6 \to A_5 \to B_5 \to A_4 \to B_7 \to A_2 \to B_0.$$

Its length is given by [comapre (12) and collect similar radicals]

(21)
$$L = 3w + 2\sqrt{v^2 + w^2} + 4\sqrt{4v^2 + w^2}$$
$$+ 3\sqrt{9v^2 + w^2} + 3\sqrt{25v^2 + w^2}.$$

In general, let the lacing have total length

(22)
$$L = \sum_{k=-n}^{n} N_k \sqrt{k^2 v^2 + w^2},$$

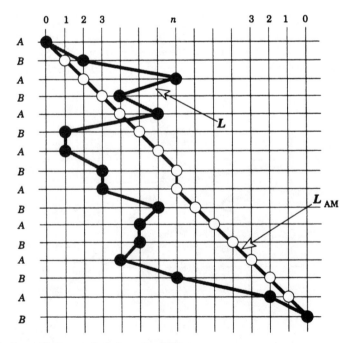

FIGURE 19.10. General lacing—reflected representations.

where clearly,

(23)
$$\sum_{k=-n}^{n} N_k = 2n + 1$$

is the net total number of downward displacements (i.e., the number of steps, since each step has a downward displacement by one lattice interval w), and

(24)
$$\sum_{k=-n}^{n} kN_k = 2n$$

is the net total number of rightward displacements by one lattice interval v. For the AM lacing, it is clear that

(25)
$$N_0 = 1, N_1 = 2n, \quad \text{all other } N_k = 0.$$

Theorem 2. *The* AM *lacing has the shortest possible total length L, and it is the unique optimum lacing.*

Proof. Let L be the reflected representation of an arbitrary lacing $\mathcal{L}$, and let L be its total length.

(i) If $N_0 \geq 1$, let us remove any one corresponding (vertical) step from $\mathcal{L}$, and let us remove the sole vertical step from $\mathcal{L}_{AM}$, rejoining the separated pieces of the representations by parallel displacement, as before. Then the two new representations, $L^\dagger$ and $L^\dagger_{AM}$, still share their end points, and both lengths are just w less than they were. Now, $L^\dagger_{AM}$ is clearly minimal, being the straight line connecting these end points. Therefore, for all $\mathcal{L}$,

(26)
$$L_{AM} \leq L.$$

(ii) Now suppose that $N_0 = 0$. This is illustrated in Figure 19.11.

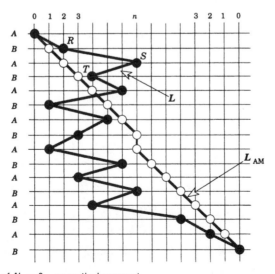

FIGURE 19.11. Case of $N_0 = 0$—no vertical segment.

It cannot be that $N_k > 0$ only for positive values of k, for then, by (23) and (24), we would have that

(27)
$$\sum_{k=1}^{n} kN_k - \sum_{k=1}^{n} N_k = N_2 + 2N_3 + \ldots + (n-1)N_n = 1,$$

which is impossible, since all $N_k \geq 0$. Therefore, there is at least one step with a negative (*leftward*) horizontal displacement, and thus there is a first leftward step ST in the downward order. It obviously *cannot* be either the *first* or the *last* step of the representation. Hence, it is preceded by a *rightward* step RS, forming an angle pointing to the right.

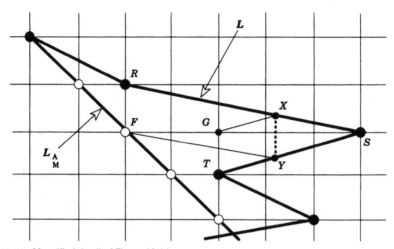

FIGURE 19.12. Magnified detail of Figure 19.11.

Now (see the enlarged detail of Fig. 19.12) let F and G be the respective lattice points in which the vertical lines through R and T meet the horizontal line through S. Then

(28)
$$|FR| = |GT| = w$$

and

(29)
$$|FS| \geq v,$$
$$\text{and}$$
$$|GS| \geq v.$$

Draw a line parallel to TS through G, and let it meet RS (as it must) at X. Now draw a vertical line (parallel to RF) through X to meet TS at Y. Clearly, $XYTG$ is a parallelogram, and therefore $|XY| = |GT| = w$, by (28).[1]

Thus we can replace the polygonal segment RST of the representation L by the polygonal segment $RXYT$, and by the triangle inequality,

(30)
$$|XY| < |XS| + |SY|,$$

---

[1]Note, too, that $XYFR$ is also a parallelogram, since the opposite sides XY and RF are equal and parallel.

so that the modified representation $L^\perp$, say, is *shorter* than L. But now $L^\perp$ *has* a vertical segment of length w; so by the same argument as in case (i), the inequality (26) prevails.

Note: The representative polygonal line $L^\perp$ is, generally, *not a representation of any lacing*, since it does not, in general, join lattice points. But this does not matter, since at this stage of the argument, we are concerned only with the *length* of the line.

We have now proved that if $\mathcal{L}_{\text{MIN}}$ is any lacing of minimal length, then it and its (horizontally reflected) representation L_{MIN} will have a total length equal to that of the AM lacing, that is, by (4),

(31) $$L_{\text{MIN}} = L_{\text{AM}} = w + 2n\sqrt{v^2 + w^2}.$$

(iii) Finally, we prove the *uniqueness* of the optimal lacing $\mathcal{L}_{\text{MIN}}$. The arguments presented in cases (i) and (ii) show that any minimal lacing $\mathcal{L}_{\text{MIN}}$ satisfies (25); that is, its (horizontally reflected) representation L_{MIN} has $2n$ straight segments, moving diagonally down and to the right by one lattice interval and one vertical segment. However, the *position* of this vertical segment in the chain does not affect the total length L_{MIN}, as indicated in (31).

Nevertheless, since L_{MIN} is not just any lattice polygon, but the representation of a *lacing*, it *must* pass through the vertical lattice line corresponding to index n just twice (corresponding to the eyelets A_n and B_n), and this is the *only* lattice line that is *not* duplicated by the reflection transformation, since it *is* the reflection line. Therefore, since the representation moves monotonically right (i.e., never to the left), the solitary vertical segment is constrained to be precisely in the index n position, as in L_{AM}. This completes the proof of Theorem 2.

REFERENCE

1. The Shoelace Problem, Department of Computer Science Technical Report No. 92-032 (1992), University of North Carolina at Chapel Hill.

chapter 20

Triangles and Proofs

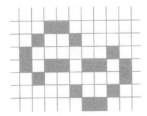

This is our third chapter concerned with the subject of triangles, but with a difference. The earlier chapters were concerned with the impact of computers on triangle geometry, whereas here we will go back to basics, taking a new look at some very classical topics.

"The point of a generalization is not just to include more cases, but also to get rid of unnecessary hypotheses, and sometimes this can lead to simpler proofs." I recall being told something like this as a graduate student by one of my first mentors, Norman Steenrod. In fact there are two kinds of generalizations. One can either weaken the hypothesis or strengthen the conclusion. The examples treated below illustrate both of these approaches. In both cases, finding the right generalization reduces what appears to be a complicated problem to essentially a mechanical verification, a matter of just "turning the crank." For our first example we are grateful once again to Donald Newman.

THE MORLEY MIRACLE

D. J. Newman

One of the sad things about the current philosophy of mathematical education is the avoidance of plane geometry.

Today's generation and perhaps their parents as well have not heard of marvels like the 9-point circle, Descartes's theorem, Ceva's theorem, or that marvel of marvels, the *Morley* triangle.

As shown in Figure 20.1, one takes an arbitrary triangle and trisects its angles, obtaining three intersection points. These form the small triangle inside the starting triangle. This small interior triangle is far from being arbitrary, however. Morley's great discovery (1899) is that it is always equilateral!

When I read, or rather tried to read, Morley's proof of this startling theorem, I found it absolutely impenetrable. I told myself that maybe in future years I would return and then understand it. I never succeeded in that, and even when I read the much simpler proof based on trigonometry or the fairly simple geometric proof due to Navansiengar, there was still too much complexity and lack of motivation. (A series of lucky breaks!) Were we to give up, forever, understanding the Morley Miracle? Or are we failing because we are asking too little? After all, Morley's theorem states that in Figure 20.1, the inner triangle always will be equilateral. The reason that all the proofs seem to be so difficult and unmotivated is probably because Morley's theorem is really only half the story. The full picture is in Figure 20.1, and tells the whole story and indeed proves itself! (This happens often in induction proofs: The fuller statement is easier to prove than the restricted one.)

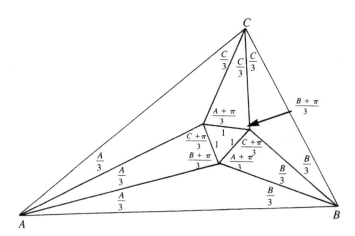

FIGURE 20.1.

So we turn to the "cheating" strategy as used in [4], namely, we start with the equilateral triangle and build out. The result is Figure 20.2. Here we have normalized matters by choosing the sides of the equilateral triangle to equal 1. Note that from symmetry it is sufficient to prove that the indicated angle is $A/3$.

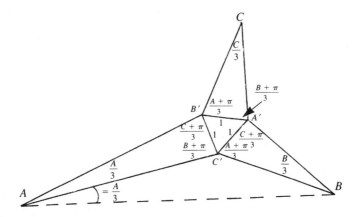

FIGURE 20.2.

At this point we could turn the proof over to a high school trigonometry student who has learned to "solve triangles." Observe that all the side lengths in Figure 20.2 are determined by angle–side–angle (ASA) of the three constructed triangles. Thus, the law of sines applied to triangle $AB'C'$ gives

$$\frac{AC'}{\sin (C + \pi)/3} = \frac{1}{\sin (A/3)},$$

so

$$AC' = \frac{\sin (C + \pi)/3}{\sin (A/3)}.$$

Similarly,

$$BC' = \frac{\sin (C + \pi)/3}{\sin (B/3)}.$$

Also

$$\angle AC'B = 2\pi - \frac{A + \pi}{3} - \frac{B + \pi}{3} - \frac{\pi}{3} = \frac{C + 2\pi}{3},$$

so for triangle $AC'B$, we have Figure 20.3.

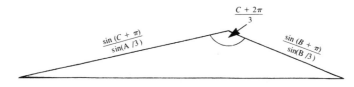

FIGURE 20.3.

For this triangle we know two sides and the included angle, (SAS), so the remaining angles are determined, and they must be $A/3$ and $B/3$, which one verifies by using the law of sines once again. QED.

REFERENCES

1. Frank Morley, Extensions of Clifford's theorem, *Amer. J. Math.* 51, 465–472 (1929).
2. J.M. Child. A Proof of Morley's Theorem, *Math. Gaz.* 171 (1922).
3. M.T. Navansiengar, *Educ. Times, New Series* 15, 47 (1909).
4. H.S.M. Coxeter, *Introduction to Geometry*, Toronto: John Wiley & Sons, 1969, pp. 23–25.

ANATOMY AND EVOLUTION OF A THEOREM ON TRIANGLES

As every schoolboy *used* to know, the medians of a triangle meet in a point (the centroid), as do the altitudes (the orthocenter). (These days you'd be lucky to find a schoolboy who even knows what a median is.)

A less familiar example of concurrence is the Fermat point. This is the point (of an acute triangle) that minimizes the sum of the distances from the three vertices. To find it, one constructs three equilateral triangles having as bases the three sides of the given triangle. Now connect the far vertex of each of these triangles to the opposite vertex of the given triangle. The intersection of these three lines is the desired point. (Fig. 20.4.)

Much less well known is the following theorem or pair of theorems related to a construction sometimes attributed to Napoleon. In Figure 20.4, choose A', B', and C' to be the centers rather than the far vertices of the three triangles. Once again the lines AA', BB', and CC' are concurrent. (The so-called Napoleon's theorem says that in this case the points A', B', and C' are themselves vertices of an equilateral triangle!) Further, the three equilateral triangles can be taken either outside or inside of the original triangle.

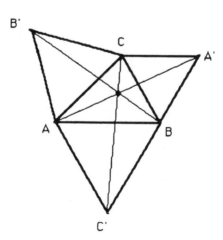

FIGURE 20.4. The Fermat Point.

First Generalization

These theorems are special cases of an infinite one-parameter family of theorems, which was first discovered over 100 years ago, although apparently it keeps being rediscovered. The theorem, attributed to Ludwig Kiepert, replaces the equilateral triangles of Fermat and Napoleon by any set of three similar isosceles triangles. If the base angle of these triangles is α, then the Fermat point corresponds to the case $\alpha = \pi/3$, Napoleon to $\alpha = \pi/6$ (plus or minus); the centroid and orthocenter are the limiting cases $\alpha = 0$ and $\alpha = \pi/2$, respectively (check it out). As α runs from $-\pi/2$ to $\pi/2$, the locus of the points of concurrence is a rectangular hyperbola known as Kiepert's hyperbola. For a recent exposition, see *Mathematics Magazine*, June 1994, 188–205.

Exercise (surprising but easy): Show that the Kiepert hyperbola also passes through the three vertices A, B, and C; thus it is the unique conic passing through the five points consisting of the centroid, the orthocenter, and the three vertices. (Do it. Get in on the fun. I dare you!)

Second Generalization

It turns out that the one-parameter family above is a special case of a triply infinite family of theorems. Referring to Figure 20.5 below, we have

Theorem 1. *Given a triangle* ABC *and points* A', B', *and* C' *such that* $\alpha = \angle BAC' = \angle B'AC$, $\beta = \angle ABC' = \angle A'BC$, *and* $\gamma = \angle A'CB = \angle ACB'$, *then* AA', BB' *and* CC' *are concurrent.*

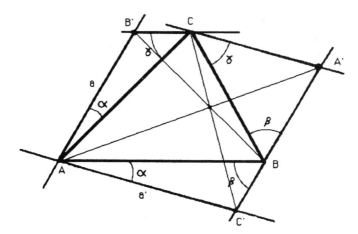

FIGURE 20.5.

Thus, the triangles on the sides of *ABC* need not be isosceles. The key condition is that the base angles of the three triangles be equal in pairs, as shown in Figure 20.5. Note that because α, β, and γ are arbitrary we have a three-parameter family of theorems; Kiepert's result is the special case where $\alpha = \beta = \gamma$. The result is so simple and natural in its statement that one suspects it must have been noted long ago, but the historical trail seems to be murky. A recent reference is a "Classroom Note" by D. Kirby (*Am. Math. Monthly*, January 1980, 45–47).

Third Generalization

The rest of the story is based on observations communicated by Clifford Gardner.

First some terminology: The lines *a* and *a'* in Figure 20.5 are called *isogonal lines* with respect to vertex *A*, meaning that they are mirror images of each other with respect to the angle bisector at *A*. Analogously, instead of reflecting in the angle bisector, one can "reflect" in the medians of the triangle. More precisely, two lines through a vertex are called *isotomic* if they meet the opposite side in points equidistant from its midpoint. The analogue of Theorem 1 also holds if the lines through the vertices are isotomic rather than isogonal. In fact, the result is true for *any* triple of concurrent lines through the vertices *A*, *B*, and *C* in the following form.

Theorem 2. *Let p, q, and r be concurrent lines through the vertices A, B, and C, respectively, of triangle ABC. Let P_A be the pencil of lines at A, and let T_A be the (unique) projective mapping on P_A that*

(1) interchanges lines AB and AC;
(2) leaves p fixed.

Define P_B, P_C and T_B, T_C similarly. For any line a in P_A, let $a' = T_A(a)$. Similarly, for b in P_B and c in P_C, let $b' = T_B(b)$, $c' = T_C(c)$. Let $C' = a' \cap b$, $B' = b' \cap c$, and $A' = c' \cap a$.

Then AA', BB', and CC' are concurrent.

Note that we now have a *five*-parameter infinity of theorems, for the coordinates of the point *P* can be chosen arbitrarily.

Actually, Theorem 2 is an immediate consequence of Theorem 1, using the well-known fact that a projectivity of the plane can be defined arbitrarily on any four independent points. Thus, in particular, Figure 20.5 becomes Figure 20.6 by leaving vertices

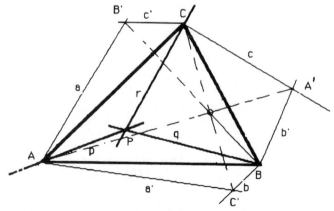

FIGURE 20.6.

A, *B*, and *C* fixed and taking the in-center into any point *P*. The usefulness of the general-ization, therefore, is not to include more cases but rather to simplify the proof. Namely, we make one more projective transformation, which carries *C* to the origin, *A* to the point at infinity on the *y*-axis, *B* to the point at infinity on the *x*-axis, and *P* to the point (1, 1) as shown in Figure 20.7.

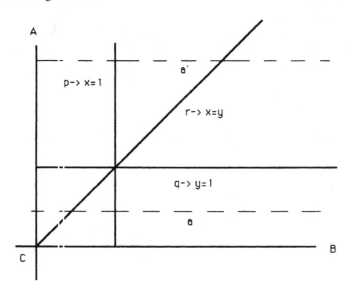

FIGURE 20.7.

Then the lines *p*, *q*, and *r* become the lines $x = 1$, $y = 1$, and $x = y$, respectively, and we have

P_A is the set of all vertical lines,
P_B is the set of all horizontal lines,
P_C is the set of all lines through the origin.

Now recall that any one-dimensional projective transformation is of the form $y = (ax + b)/(cx + d)$, so T_A maps the lines $x = a$ to the lines $x = 1/a$. This is the unique projectivity that interchanges the *y*-axis and the line at infinity, leaving the line $x = 1$ fixed. We will denote the vertical line whose equation is $x = a$ simply by *a*, the horizontal line $y = b$ by *b*, and the line $y = cx$ through the origin by *c*. Then with similar definitions of T_B and T_C, we have

$$a' = T_A(a) = \frac{1}{a}, \quad b' = T_B(b) = \frac{1}{b}, \quad c' = T_C(c) = \frac{1}{c}.$$

It is now a matter of simple high school analytic geometry to calculate the coordinates of the points *A'*, *B'*, and *C'* and the equations of the lines *AA'*, *BB'*, and *CC'* and verify that they are concurrent, namely,

$$C' = a' \cap b = (1/a, b), \quad A' = b' \cap c = (1/bc, 1/b), \quad B' = c' \cap a = (a, a/c),$$

thus, *AA'* has equation $x = 1/bc$, *BB'* has equation $y = a/c$, and *CC'* has equation $y = (ab)x$. Then $AA' \cap BB' = (1/bc, a/c)$, which lies on the line *CC'*.

Fourth (and Last) Generalization

Observe that Theorem 2 is a theorem of projective geometry. This implies that Theorem 1 is "absolute," meaning that it holds in elliptic and hyperbolic as well as Euclidean geometry (in the hyperbolic case one makes the hypothesis that the points A', B', and C' exist). This is not obvious even for the simplest special case, namely, the concurrence of the medians, which was the starting point of this exposition. The usual Euclidean proof makes strong use of the parallel postulate, as one needs the line connecting the midpoints of two sides of a triangle to be *parallel* to the third side. It can be shown using other methods that the median theorem also holds in the non-Euclidean cases. By contrast, theorem 2 covers all three cases directly.

Polyominoes

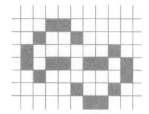

TILING RECTANGLES WITH POLYOMINOES
Solomon W. Golomb

In 1900, when the great German mathematician David Hilbert, in an address to the International Congress of Mathematicians assembled in Paris, laid out his agenda for twentieth-century mathematics, his "most wanted list" of twenty-three unsolved problems included, as number 18, a question concerning the ways in which n-dimensional Euclidean space (including $n = 2$ and $n = 3$) can be "tiled" or "packed" with congruent copies of a single geometric figure. Specifically, he asked:

> #1. "Is there in n-dimensional Euclidean space ... only a finite number of different kinds of groups of motions with a [compact] fundamental region?"

> #2. "Whether polyhedra also exist that do not appear as fundamental regions of groups of motions, by means of which nevertheless by a suitable juxtaposition of congruent copies a complete filling up of all [Euclidean] space is possible?"

The groups of motions (#1) were identified by L. Bieberbach, but examples answering #2 in the affirmative were found in 3 dimensions (K. Reinhardt, 1928) and in 2 dimensions (Heesch, 1935). Simpler, related, and more general examples were subsequently found by R. M. Robinson, S. K. Stein, and others. An example related to Heesch's is shown in Figure 21.1. (Smaller examples exist, but it is somewhat harder to prove that they have the required property.) It is not hard to show that the only way this figure can tile the plane is as shown in Figure 21.2. However, it is not a "fundamental domain," for the unique motion that carries A onto B combines a reflection about the dotted line with an upward translation by 2 units, but the image of B under this motion is not a tile.

Hilbert never anticipated Gödel's incompleteness theorem, let alone the possibility that some of his problems would be proved "undecidable." Just as his tenth problem, involving finding all integer solutions of given ("diophantine") equations was shown (by Julia Robinson, Ju. V. Matijasevic, *et al.*) to be computationally undecidable, so too the general question of whether it is possible to tile the plane with congruent copies of a given finite set of tiles was proved undecidable by Hao Wang. (Wang showed the equivalence of the tiling problem to the "halting problem" for Turing machines, a standard undecidable problem, on the one hand, and to the undecidability of the class of logical formulas containing three quantifiers, the so-called AEA-formulas, on the other.) R. Berger and other students of Wang extended his results, e.g., to showing that the question of whether it is possible to tile the plane (or a smaller region, such as a rectangle) using congruent copies of a *single* geometric figure is also computationally undecidable.

This chapter summarizes what is known about a particular special case of the tiling problem in two dimensions.

The only tiles we shall consider are *polyominoes,* where an "*n*-omino" is any connected figure obtained by taking *n* identical unit squares and connecting them along common edges. Thus, except for orientation, there is only one *monomino* (the unit square itself) and one *domino* (the eponymous ancestor of this entire clan). There are two distinct *trominoes,* five different *tetrominoes,* twelve *pentominoes,* thirty-five *hexominoes,* etc. The simpler ones of these are shown in Figure 21.3. (We do not regard mirror images ("reflections") as two distinct shapes.)

A typical problem in polyomino theory is to determine which polyominoes have the property that unlimited copies of a specific one tile the entire plane, tile a single quadrant of the plane, or tile an infinite strip bounded by two parallel straight lines, etc. (Figure 21.2 shows how to tile a strip, hence the plane, with a particular polyomino.) Restrictions

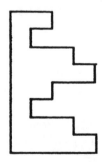

FIGURE 21.1. A figure that tiles the plane but is not a "fundamental region."

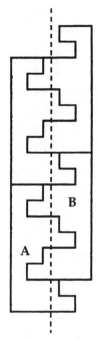

FIGURE 21.2. Tiling the plane with the shape in Figure 21.1.

can be placed on which, if any, rotations and/or reflections of the basic shape may be used in the tiling. Tilings may be studied in which two, three, or some other number of different tile shapes are allowed in the tiling. One may ask about those shapes such that several identical copies may be assembled to make an enlarged scale model, or *replica*, of the

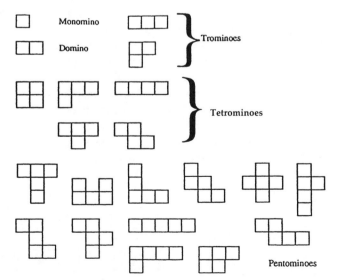

FIGURE 21.3. The simpler polyominoes.

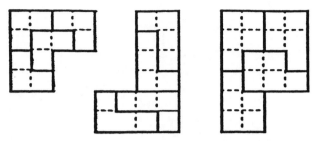

FIGURE 21.4. Three "rep-tiles": a tromino, a tetromino, and a pentomino.

original shape, as in Figure 21.4. (Many years ago, in 1962, I coined the term *rep-tiles* for these shapes.)

All of these questions have been studied. However, this chapter focuses primarily on the question of which polyomino shapes have the property that some finite number of copies of the basic shape, allowing all rotations and reflections, can be assembled to form a rectangle. The problem is challenging because no a priori limit can be established, given an arbitrary n-omino, on the minimum number of copies that *may* be required to form a rectangle. No explicit expression in n, such as n^{n^n}, grows fast enough to guarantee that when all possible arrangements of that many copies of the basic shape have been examined and no rectangle has been found, then no rectangle involving yet more copies exists. This follows from the result, mentioned earlier, that the *general* question of whether an *arbitrary* polyomino tiles a rectangle is "computationally undecidable." Fortunately, in any *specific* case, there is a high likelihood (though no certainty) of answering the question. But there is no cookbook recipe, no handbook procedure, that can be routinely applied to indicate whether a given polyomino shape tiles some (possibly huge) rectangle. And it is this lack of a general decision procedure that makes this study so interesting and challenging.

In 1968, David A. Klarner defined the *order* of a polyomino P as the minimum number of congruent copies of P that can be assembled (allowing translation, rotation, and reflection) to form a rectangle. For those polyominoes that do not tile any rectangle, the order is undefined. A polyomino has order 1 if and only if it is itself a rectangle.

A polyomino has order 2 if and only if it is "half a rectangle," since two identical copies of it must form a rectangle. In practice, this means that the two copies are 180° rotations of each other when forming a rectangle. Some examples are shown in Figure 21.5.

There are no polyominoes of order 3. (This was proved by Ian Steward of the University of Warwick, England.) In fact, the only way any rectangle can be divided up into three identical copies of a "well-behaved" geometric figure is to partition it into three *rectangles* (see Figure 21.6), and by definition a rectangle has order 1.

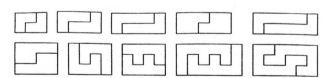

FIGURE 21.5. Some polyominoes of order 2.

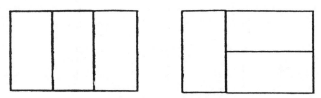

FIGURE 21.6. How three identical rectangles can form a rectangle.

There are various ways in which four identical polyominoes can be combined to form a rectangle. One way, illustrated in Figure 21.7, is to have four 90° rotations of a single shape forming a square.

Another way to combine four identical shapes to form a rectangle uses the fourfold symmetry of the rectangle itself: left–right, up–down, and 180° rotational symmetry. Some examples of this appear in Figure 21.8.

There are also more complicated order-4 patterns that were found by Klarner, two of which are illustrated in Figure 21.9.

Beyond order 4, there is a systematic construction found by Golomb in 1985 that gives examples of order $4s$ for every positive integer s. Eleven isolated examples of small polyominoes with respective orders 10, 18, 24, 28, 50, 76, 92, 96, 138, 192, and 312 are known.

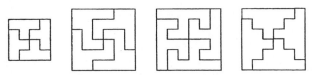

FIGURE 21.7. Polyominoes of order 4 under 90° rotation.

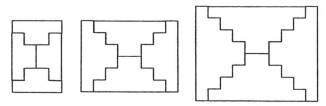

FIGURE 21.8. Polyominoes of order 4 under rectangular symmetry.

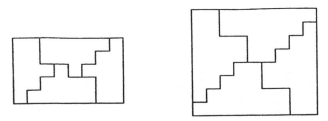

FIGURE 21.9. Another order-4 construction by Klarner.

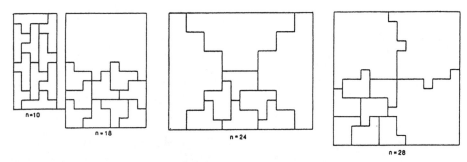

FIGURE 21.10. Four "sporadic" polyominoes, of orders 10, 18, 24, and 28, respectively.

Figure 21.10 shows the isolated examples of order 10 (Golomb, 1966) and orders 18, 24, and 28 (Klarner, 1969).

Figure 21.11 shows the example of order 50, found by William Rex Marshall of Dunedin, New Zealand, in 1990.

Figure 21.12 shows the examples of orders 76 and 92, both found by Karl A. Dahlke in 1987. I had mentioned these two problems in my talk at the Strens Memorial Conference on Recreational Mathematics (Calgary, 1986), and Ivars Peterson included them in *Science News* in his article about the Strens Conference. When Dahlke sent me solutions, which he said he found using only a personal computer, I notified Peterson, who interviewed Dahlke and learned that he is totally blind. I later heard that these two tilings had actually been discovered earlier, by T. W. Marlow in 1985.

The heptomino of order 76 in Figure 21.12 cannot tile its minimum rectangle with 180° rotational symmetry. This is also true of the dekomino in Figure 21.13 of order 96, whose minimum rectangle (the 30 × 32) was discovered by William Rex Marshall in 1991. In 1995, Marshall also found the minimum rectangles for the order 192 octomino (32 × 48) and the order 138 dekomino (30 × 46). These will be considered later.

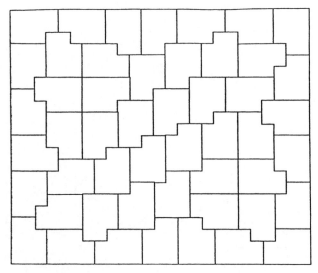

FIGURE 21.11. An 11-omino of order 50.

FIGURE 21.12. A heptomino of order 76 and a hexomino of order 92.

Finally, Figure 21.14 shows the example of order 312 (Dahlke, 1988), although in this case it is not absolutely certain that no smaller number of copies of the octomino in question form a rectangle.

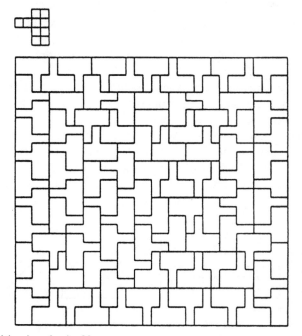

FIGURE 21.13. A dekomino of order 96.

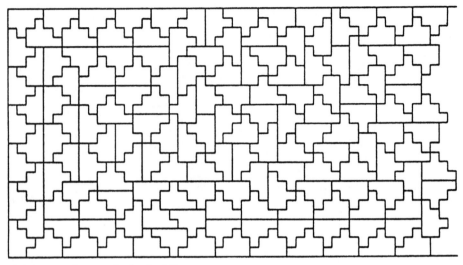

FIGURE 21.14. An example of order 312.

No polyomino whose order is an odd number greater than 1 has ever been found, but the possibility that such polyominoes exist (with orders greater than 3) has not been ruled out.

The known *even* orders of polyominoes are all the multiples of 4, as well as the numbers 2, 10, 18, 50, and 138. Curiously, those even orders that are not multiples of 4 all exceed multiples of 8 by two. Whether there are other even orders, and what they might be, is still unknown. The smallest even order for which no example is known is order 6. Figure 21.15 shows one way in which six copies of a polyomino are fitted together to form a rectangle, but the polyomino in question (as shown) actually has order 2. Michael Reid recently found a *heptabolo* (a figure made of seven congruent isosceles right triangles) of order 6, also shown in Figure 21.15.

The Golomb construction for polyominoes of order 4s gives its first new example, order 8, when $s = 2$. The underlying tiling concept of how to fit 8 congruent shapes together to form a rectangle is shown in Figure 21.16.

Although the shape used in Figure 21.16 is not a polyomino, the same concept is realized with the 12-omino shown in Figure 21.17. (The shape used in Figure 16 is a *triabolo!*)

To show that there are infinitely many dissimilar polyominoes having order 8, we form the family of polyominoes shown in Figure 21.18. For each integer $r \geq 1$, this con-

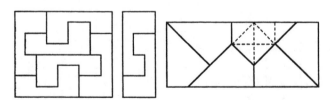

FIGURE 21.15. A 12-omino of order 2 that suggests an order-6 tiling, and Michael Reid's order-6 "heptabolo." (Is there any *polyomino* of order 6?)

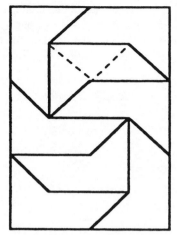

FIGURE 21.16. A rectangle formed from eight congruent pieces.

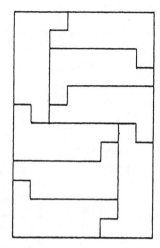

FIGURE 21.17. A polyomino of order 8.

struction produces a $(3r^2 + 6r + 3)$-omino of order 8, and clearly no two of these are similar.

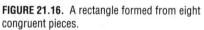

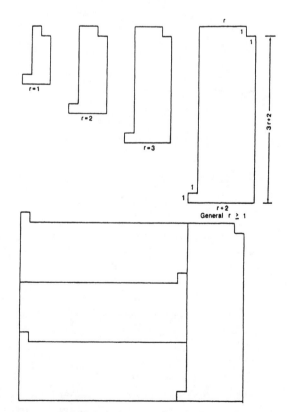

FIGURE 21.18. Dissimilar polyominoes of order 8 and how to stack them.

It is also easy to show that none of these polyominoes can have order less than 8. The proof begins by observing that only the "heel of the boot" can be in any *corner* of the rectangle to be tiled. Then the "toe" of the boot must be mated with the notch at the top-back of another boot. The quickest way to finish off the rectangle then requires eight copies of the polyomino.

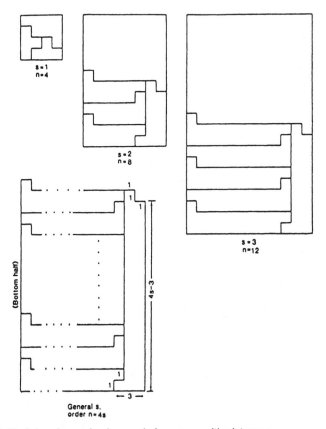

FIGURE 21.19. Polyominoes of order $n = 4s$ for every positive integer s.

In Figure 21.19, we see a construction for a polyomino of order $n = 4s$ for every $s = 1, 2, 3, 4, \ldots$. (Starting with a $2 \times (4s - 2)$ rectangle, we remove a single square from one corner and attach it as a "toe" at the opposite corner to obtain the polyomino of order $4s$.)

The idea shown in Figure 21.18 can be applied not only to order $n = 8$, but to any order $n = 4s$, to obtain infinitely many dissimilar polyominoes of order $4s$. The general construction involving both r and s begins with a rectangle that is $(r + 1) \times (2s - 1)$ $(r + 1)$ and moves a single 1×1 square from the top-back of the "boot" to become a "toe" at the opposite corner. (The proof that the resulting figure truly has order $n = 4s$ is analogous to the proof given for $n = 8$.)

As the reader is probably aware by now, the game polyominologists play is this: given a polyomino, will it or won't it tile? In recent years, whenever I have publicized a specific polyomino whose ability to tile any rectangle was not yet decided, someone with a good

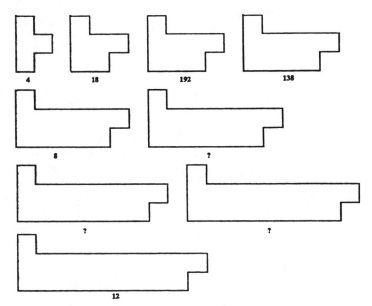

FIGURE 21.20. Infinite family of polyominoes. Does each one tile a rectangle? (The number below each figure is its order, if known.)

computer program has come forth, usually within a year, with a rectangle-tiling solution. This time, I suggest an infinite family of polyominoes, the first several of which are known to tile rectangles, as is every fourth one throughout the family. Can you show that *any* or *all* of the others can tile rectangles? The family is illustrated in Figure 21.20. Two of the minimum rectangles, discovered in 1995 by William Rex Marshall, are shown in Figure 21.21.

FIGURE 21.21. An octomino of order 192 and a dekomino of order 138.

We now return to the tiling of figures other than rectangles. If a polyomino has no order (i.e., if it cannot tile any rectangle), it may still be able to tile the entire plane, or various subregions of the plane, such as an infinite strip or a bent strip. Such tilings are illustrated in Figure 21.22, using the X-pentomino, the F-pentomino, and the N-pentomino, respectively.

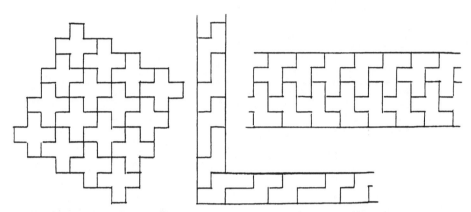

FIGURE 21.22. The X-pentomino tiles the plane; the F-pentomino tiles a strip; the N-pentomino tiles a bent strip.

In Figure 21.23, we see a tiling hierarchy for polyominoes (Golomb, 1966). A polyomino that can tile any of the regions specified in the hierarchy can also tile all the regions lower in the hierarchy. Thus, the "true category" of a polyomino is the *highest* box in the hierarchy that it can occupy. Most of this chapter is concerned with polyominoes that occupy the highest box—i.e., they can tile rectangles. The "true categories" that are known to have members are Rectangle, Bent Strip, Strip, Itself, Plane, and Nothing. Figure 21.24 shows an example of the category "Nothing." (The others have already been illustrated. The rep-tiles in Figure 21.4 illustrate the category "Itself.") For each of the other positions in the hierarchy, it is an open question whether any polyomino has that position as its "true category."

Most of the inclusion relations in Figure 21.23 are immediately obvious. To see that a polyomino that tiles a bent strip tiles both a quadrant and a straight strip, we first observe that bent strips (as in Figure 21.22) can be "nested" to cover a quadrant of the plane. We will show how to go from the bent strip to the straight strip after taking up one last important subject.

The most challenging unsolved problem about planar tiling involves shapes that can tile the infinite plane, but only nonperiodically. R. M. Robinson was the first to find a finite (but very large) set of shapes such that unlimited copies of members of the set could be used to tile the infinite plane, but only nonperiodically. He eventually reduced the size of the set to six. Independently, Roger Penrose found a set of six shapes with the nonperiodic-only tiling property, which he eventually reduced to a set of *two* shapes. I believe that a year-2000 version of Hilbert's list would include the question, Is there a *single* geometric shape S such that congruent copies of S can be used to tile the infinite plane, but only nonperiodically?

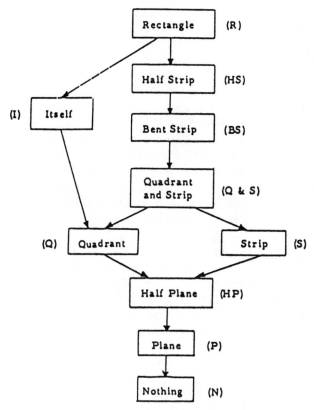

FIGURE 21.23. The hierarchy of tiling capabilities for polyominoes.

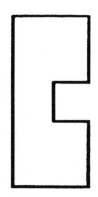

FIGURE 21.24. A 9-omino that cannot tile the plane.

To understand this question better, note that the answer is negative if we consider an infinite strip instead of the plane, even if we allow more than one tile shape to be used. To see this, consider a horizontal strip. Now choose any vertex on the upper edge and trace a "jagged edge path" by the following rules: If there is a downward edge, follow it. If not, move to the right. Continue to prefer "downward" and secondarily, "rightward." Sometimes one may even be forced to go "upward" or "leftward" to trace around a protuberance on a tile, but clearly one will eventually arrive at the bottom edge of the strip, and one can easily obtain a uniform bound on the number of "moves" required in terms of the thickness of the strip and the sizes and shapes of the tiles. This means that there is only a finite number of possible jagged paths, so there must be two different starting vertices giving the same jagged path. But then, if one takes copies of the region between these two identical paths and lays them end to end, one gets the desired periodic tiling of the strip. Further, we see, as promised, that polyominoes that tile a bent strip also tile a straight strip (applying the same argument to either one of the semi-infinite arms of the bent strip), and in fact the tiling can be made periodic.

In Figure 21.25, we see how the "P-pentomino" is used to tile a quadrant nonperiodically, by iterating the rep-tile subdivision of the P-pentomino into smaller copies of itself. We enlarge the picture, repeat the process, enlarge the picture, repeat the process, and "eventually" we have filled the first quadrant of the plane. Reflections in the coordinate axes are used to cover the entire plane nonperiodically. (This does not solve the previously mentioned problem, because the P-pentomino can also tile the plane periodically.)

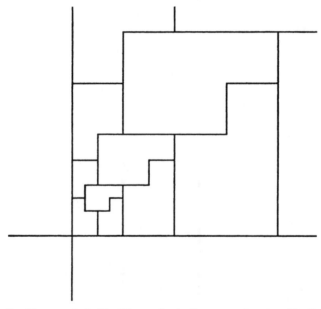

FIGURE 21.25. Rep-tilic nonperiodic tiling of a quadrant with congruent copies of the P-pentomino.

Not only the P-pentomino, but *every* polyomino rep-tile can tile a quadrant. To show this, let J be any polyomino rep-tile, and let R be the smallest rectangle containing J with sides parallel to the gridlines of J. We first prove the lemma that J must cover at least one of the four corners of its minimum rectangle R. Suppose not. In Figure 21.26, we see a polyomino K that occupies none of the corners of its minimum rectangle. Consider the leftmost square of K along the bottom of its minimum rectangle, indicated with an asterisk. If K were a rep-tile, we could divide it into congruent replicas, and we could iterate this process until each replica was so small that it would fit within a single square of the original grid. At this stage, consider the replica k of K covering the lower left corner of the square "*". It fits entirely within that square and fills a corner of it, so clearly k occupies a corner of its minimum rectangle. But the same could be true of K.

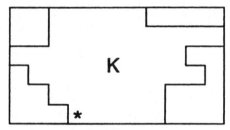

FIGURE 21.26. A polyomino inscribed in its minimum rectangle.

Now, knowing that J must occupy at least one corner of its minimum rectangle, put such a corner square of J into the lower left corner of the first quadrant of the plane. Now, perform the rep-tilic subdivision of J, enlarge back to the original scale, and continue to subdivide and enlarge (as we did with the P-pentomino in Figure 21.25) until we have filled up the entire first quadrant. (We do not need to invoke König's lemma to assert that a limiting tiling exists. The method is constructive. Given any radius R, however large, we can specify the tiling of the first quadrant out to distance R from the origin, in such a way that this tiling does not change as R is increased. This may involve looking only at every mth iteration of the rep-tilic subdivision, if the J nearest the origin is moved by a single iteration.)

The proofs that a polyomino that tiles a bent strip will tile a straight strip and that a polyomino that tiles itself (i.e., a rep-tile) will tile a quadrant were given in Golomb (1966).

Finally, returning to the subject of nonperiodic tilings of the plane, the natural question to ask in our present context is whether there exist "nonperiodizable" tilings using polyominoes. The best current result is the set of *three* shapes, shown in Figure 21.27, which can be used to tile the plane, but not periodically. (This set was recently discovered by Penrose, who writes that he derived these from a tiling set by Robert Ammann.) Penrose's aperiodic tilings also have connections to "quasicrystals," which have been of major interest to chemists in recent years.

Arguments that show that certain sets of "pieces" tile the plane, but not periodically, are both clever and subtle (see Grünbaum and Shepard, 1987, Chapter 10), and undoubt-

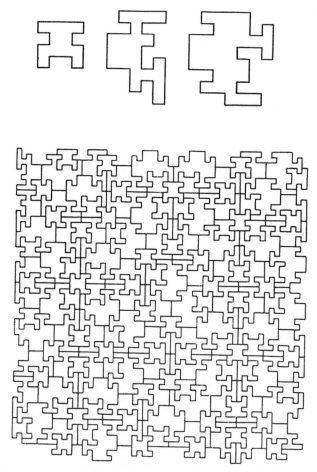

FIGURE 21.27. Roger Penrose's set of three polyominoes that can be used to tile the plane, but not periodically.

edly lie outside the range of anything Hilbert ever contemplated. But Hilbert's insight, which led him to include a problem about tilings and packings in his famous List, was profound. This is a subject that is accessible to amateurs but lies close to the very heart of mathematics and continues to provide a seemingly inexhaustible supply of intriguing and provocative questions.

Note: This material is based on Chapter 8 of *Polyominoes—Revised Edition*, by Solomon W. Golomb, Princeton University Press, 1994. It is one of the new chapters added to the text of the original edition of *Polyominoes*, published in 1965 by Charles Scribner's Sons.

REFERENCES

Robert Berger, The undecidability of the domino problem, *Memoirs of the American Math. Society* 66, 1–72 (1966).

Karl A. Dahlke, The Y-hexomino has order 92, *Journal of Combinatorial Theory, Series A* 51, 125–126 (1989).

Karl A. Dahlke, A heptomino of order 76, *Journal of Combinatorial Theory, Series A* 127–128 (1989).

Karl A. Dahlke, Solomon W. Golomb and Herbert Taylor, An octomino of high order, *Journal of Combinatorial Theory, Series A* 70, 157–158 (1995).

Martin Gardner, Mathematical Games: On "rep-tiles," polygons that can make larger and smaller copies of themselves, *Scientific American* 208, 154–164 (1963).

Solomon W. Golomb, Replicating figures in the plane, *Mathematical Gazette* 48, 403–412 (1964).

Solomon W. Golomb, Tiling with polyominoes, *Journal of Combinatorial Theory* 1, 280–296 (1966).

Solomon W. Golomb, Tiling with sets of polyominoes, *Journal of Combinatorial Theory* 9, 60–71 (1970).

Solomon W. Golomb, Polyominoes which tile rectangles, *Journal of Combinatorial Theory, Series A* 51, 117–124 (1989).

B. Grünbaum and G.C. Shephard, *Tilings and Patterns,* Freeman, New York (1987).

A.S. Kahr, E.F. Moore, and H. Wang, Entscheidungsproblem reduced to the ∀∃∀ case. *Proceedings National Academy of Sciences USA* 48, 365–377, 1962.

David A. Klarner, Packing a rectangle with congruent N-ominoes, *Journal of Combinatorial Theory* 7, 107–115 (1969).

William Rex Marshall, private communications dated 14 May, 1990, 25 November, 1991, and 6 June, 1995.

Roger Penrose, *Shadows of the Mind,* Oxford University Press, 1994.

Michael Reid, private communications from 1992 to 1995.

Ian Stewart and A. Wormstein, Polyominoes of order 3 do not exist, *Journal of Combinatorial Theory, Series A* 61, 130–136, 1992.

chapter 22

A Pattern Problem,
A Probability Paradox,
and A Pretty Proof

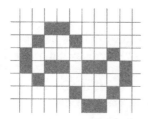

PROBLEMS AND PATTERNS

If S is a finite set of nonnegative integers, we define mex$\{S\}$ as the smallest nonnegative integer that is not a member of S. (mex is an abbreviation for *minimum-excluded* (Berlekamp, Conway, and Guy, *Winning Ways*).)

The two-dimensional array A below, assumed to extend out to infinity, satisfies the following condition:

(*) *The ijth entry $a_{ij} = $ mex$\{S_{ij}\}$, where S_{ij} is the set of entries to the left of it in row i or above it in column j.*

Problem: *What is the entry in row 777, column 1001?*
[Of course we are really asking for an explicit expression for the ijth entry.]

It is clear that there is one and only one array satisfying (*), namely, a_{00} must be zero. Hence, entries a_{01} and a_{10} are 1, etc., and one sees that in building up the matrix, row by row (or any other way), the entries at each stage are "forced," and one can grind out terms at will without resorting to computational technology. Here as elsewhere we index rows and columns staring with 0. What emerges is highly unexpected. Indeed, it is hard to see how one could have guessed the result without this preliminary experimentation. Table 22.1 is a section of the array.

What is going on here? The partitioning of A in Table 22.2 shows what is happening.

As noted, a_{00} must be 0, a_{01} and a_{10} are 1, and a_{11} is again 0. If we call this 2×2 matrix A_{11}, then we see that the adjacent 2×2 matrices A_{12} and A_{21} to the right and below A_{11} are A_{11} with 2 added to each of the entries, and A_{22} is A_{11} translated along the diagonal. We now have a 4×4 matrix A_{44}. To get the 4×4 matrices below and to the right of A_{44}, one adds 4 to the entries of A_{44}, and then A_{44} is repeated on the diagonal, giving an 8×8 array. Continue in this way to get a 16×16 array, and so on. The general structure is now apparent.

The remarkable feature of the array is the totally unexpected role played by the powers of 2. There is nothing in condition (*) that would lead one to anticipate this. Having noted it, however, one then naturally writes the integers in binary notation, at which point the secret is revealed. Table 22.3 is an initial section of the array.

TABLE 22.1.

0	1	2	3	4	5	6	7	8	9	10	11	12	13	14	15	16	17	18
1	0	3	2	5	4	7	6	9	8	11	10	13	12	15	14	17	16	19
2	3	0	1	6	7	4	5	10	11	8	9	14	15	12	13	18		
3	2	1	0	7	6	5	4	11	10	9	8	15	14	13	12			
4	5	6	7	0	1	2	3	12	13	14	15	8	9	10	11			
5	4	7	6	1	0	3	2	13	12	15	14	9	8	11	10			
6	7	5	4	2	3	0	1	14	15	12	13	10	11	8	9			
7	6	4	5	3	2	1	0	15	14	13	12	11	10	9	8			
8	9	10	11	12	13	14	15	0	1	2	3	4	5	6	7			
9	8	11	10	13	12	15	14	1	0	3	2	5	4	7	6			
10	11	8	9	14	15	12	13	2	3	0	1	6	7	4	5			
11	10	9	8	15	14	13	12	3	2	1	0	7	6	5	4			
12	13	14	15	8	9	10	11	4	5	6	7	0	1	2	3			
13	12	15	14	9	8	11	10	5	4	7	6	1	0	3	2			
14	15	12	13	10	11	8	9	6	7	4	5	2	3	0	1			
15	14	13	12	11	10	9	8	7	6	5	4	3	2	1	0			
16	17	18																
17	18																	

TABLE 22.2.

0	1	2	3	4	5	6	7	8	9	10	11	12	13	14	15	16	17	18
1	0	3	2	5	4	7	6	9	8	11	10	13	12	15	14	17	16	19
2	3	0	1	6	7	4	5	10	11	8	9	14	15	12	13	18		
3	2	1	0	7	6	5	4	11	10	9	8	15	14	13	12			
4	5	6	7	0	1	2	3	12	13	14	15	8	9	10	11			
5	4	7	6	1	0	3	2	13	12	15	14	9	8	11	10			
6	7	5	4	2	3	0	1	14	15	12	13	10	11	8	9			
7	6	4	5	3	2	1	0	15	14	13	12	11	10	9	8			
8	9	10	11	12	13	14	15	0	1	2	3	4	5	6	7			
9	8	11	10	13	12	15	14	1	0	3	2	5	4	7	6			
10	11	8	9	14	15	12	13	2	3	0	1	6	7	4	5			
11	10	9	8	15	14	13	12	3	2	1	0	7	6	5	4			
12	13	14	15	8	9	10	11	4	5	6	7	0	1	2	3			
13	12	15	14	9	8	11	10	5	4	7	6	1	0	3	2			
14	15	12	13	10	11	8	9	6	7	4	5	2	3	0	1			
15	14	13	12	11	10	9	8	7	6	5	4	3	2	1	0			
16	17	18																
17	18																	

TABLE 22.3.

0	1	10	11	100	101	110	111	1000
1	0	11	10	101	100	111	110	1001
10	11	0	1	110	111	100	101	1010
11	10	1	0	111	110	101	100	1011
100	101	110	111	0	1	10	....	

Definition: If i and j are integers, we consider their binary expansions as vectors over $\mathbf{Z}_2$. Then their *nim sum*, $k = i \oplus j$, is the integer whose binary expansion is the vector sum of i and j. Thus, the rth digit of k is 1 if and only if 1 is the rth digit of i or j but not both.

We adopt the convention that rows and columns of A are labeled starting from 0 rather than 1. The story is then the following:

Theorem. *The entry a_{ij} in A is $i \oplus j$.*

An interesting consequence of the theorem is that A is the multiplication table for an abelian group structure on the nonnegative integers, for the nim sum treats an n-digit binary expansion of an integer as an element of the n-dimensional vector space over $\mathbf{Z}_2$. One can verify that the entries of A satisfy the associative law.

Tables 22.1–22.3 supply very convincing "evidence" for the correctness of our theorem, but the proof below is perhaps not completely trivial.

Proof: Let $i \oplus j = n$. We must show that if m is less than n, then either $m = i' \oplus j$ with $i' < i$, or $m = i \oplus j'$ with $j' < j$.

If d is a binary vector, let d_k be its kth digit (starting from the right). Now let r be the largest subscript such that $m_r \neq n_r$. Then $n_r = 1$, and $m_r = 0$, since $m < n$, so either i_r or j_r is 1, say $i_r = 1$. Let i' be the (unique) binary vector such that $i' \oplus j = m$. Then $i'_k = i_k$ for $k > r$, and $i'_r = 0$. So i' is less than i; thus, $m = a_{i'j}$ *is to the left of* a_{ij}.

What about the original question, to find the entry in row 777, column 1001? Well, we have

$$777 \rightarrow 1\ 1\ 0\ 0\ 0\ 0\ 1\ 0\ 0\ 1 \quad \text{and}$$

$$1001 \rightarrow 1\ 1\ 1\ 1\ 1\ 0\ 1\ 0\ 0\ 1; \quad \text{so}$$

$$777 \oplus 1001 = \ 0\ 0\ 1\ 1\ 1\ 0\ 0\ 0\ 0\ 0 = 224.$$

Remarks: The theorem survives intact for n-dimensional arrays; namely, (*) becomes the condition that the entry at $(i_1, \ldots, i_k \ldots, i_n)$ is

mex{entries at $(i_1, \ldots, i'_k, \ldots, i_n)$ for all k, where $i'_k < i_k$}.

Then entry $(i_1, \ldots, i_k \ldots, i_n)$ turns out to be $i_1 \oplus i_2 \oplus \cdots \oplus i_n$.

Why the term "nim sum"? All right, think of a game of Nim with n piles, where pile k contains i_k elements. People familiar with Nim will know that a *position* is a second-player win if and only if the nim sum of the i_k is 0. In general, we may think of $i_1 \oplus i_2 \oplus \cdots \oplus i_n$ as the *value* of the Nim position $(i_1, \ldots, i_k \ldots, i_n)$, and our theorem asserts that a player can move to any lower value by withdrawing from only one of the piles. Two-pile Nim is of course not interesting, because the 0s occur exactly on the diagonal; but in the three-pile case, for example, $1 \oplus 2 \oplus 3 = 0$.

One can create variations of the original problem by changing the definition of the set S_{ij}. One interesting variation is to include in S the backward diagonals, that is, augment S_{ij} to include all $a_{i'j'}$ with $i' < i$, $j' < j$, and $i' - j' = i - j$. For this case no explicit expression for the a_{ij} is known, but there is a method for finding the 0s. This corresponds to a 2-pile Nim-like game in which a move consists in reducing one of the piles by an arbitrary amount, as in ordinary Nim, or reducing both piles by the same amount.

A Nim variation that is still unsolved is ordinary Nim with the additional possibility that once during the game either player, but not both, may "pass," except that a pass is not allowed after all the pieces have been removed. Again, the last player to move is the winner.

ANOTHER PROBABILITY PARADOX

A random variable X with a known distribution takes on integer values from a finite ordered set S. For example, S might be the integers from 1 to 10. Let p_i denote the probability that X equals i, and let D_i denote the probability that X is greater than or equal to i.

The "game" (against nature) Γ^2 consist in making a sequence of independent observations of values $x_1, x_2, \ldots$ of X, where the sequence terminates as soon as some x_i is the "second highest," meaning that there is exactly one previous term of the sequence that is greater than or equal to x_i. The observer then receives a *payoff* equal to this x_i.

The game Γ^k is the same except that second highest is replaced by the kth highest.

The game Γ_k is the same as Γ^k except that the kth highest is replaced by the kth lowest.

Example: The payoff to Γ^4 of the sequence $(3, 7, 2, 5, 1, 6, 1, 5^*, 9, 2, \ldots)$ is 5, since exactly three previous observations, $x_2, x_4,$ and x_6, were 5 or greater.

Note: In the games Γ_1 and Γ^1, the payoff is simply x_1.

QUESTION
Which of the games $\Gamma^2, \Gamma^{17}, \Gamma_2$, etc., is most advantageous for the observer?

ANSWER
They are all equivalent.

Theorem. *The probability distributions of the payoffs of Γ^k and Γ_k are the same as the original distribution on S.*

This statement seems at first counterintuitive. One feels that it ought to be better to bet on the second best rather than the seventeenth best or second worst.

This theorem is essentially the discrete case of a result in probability theory known as Ignatov's theorem. The result can be proved by direct computation, looking at the combinatorial possibilities among the set of all sequences of observations and observing some identities among binomial coefficients. Instead, I present a short indirect proof that requires essentially no computation. The informal argument that follows can be easily formalized. It depends on a tricky characterization of the payoff of Γ^k.

Notation: given a sequence $Z = (x_1, x_2, \ldots)$ of observations, let Z_r be the subsequence consisting of all terms of Z that are greater than or equal to r.

Terminology: We will call the payoff of Γ^k the *k-payoff*.

Lemma. *The k-payoff of the sequence $Z = (x_1, x_2, \ldots)$ is the smallest number r in S such that the kth term of Z_r is r.*

Proof: If the game terminates with payoff r, this means that exactly $(k-1)$ previous observations were greater than or equal to r; but this is means that r is the kth term of Z_r.

Finally, r must be the smallest number with this property, for if there were a smaller one, the game would have terminated earlier. ∎

The theorem follows directly from the lemma, namely, the probability that the kth term of Z_s is [is not] s is p_s/D_s $[D_{s+1}/D_s]$. Further, if the kth term of Z_s is not s, then it is greater than s, and this gives no information on the value of the kth term of Z_{s+1}. Thus, the probability of payoff r is $D_1(D_2/D_1) \cdots (D_r/D_{r-1})(p_r/D_r) = p_r$.

The fact that the theorem also holds for the games Γ_k follows by symmetry.

But one can say more than this. Given a sequence Z, suppose that one knows that the 2-payoff is, say, 5. What can one infer from this about the 3-payoff? The 17-payoff? Answer: nothing. In other words, given the sequence Z, if j is unequal to k, then the j- and k-payoffs are *independent*. This follows again from the lemma, which shows that the j-[k]-payoff depends only on the jth [kth] terms of the sequences $Z_1, Z_2, \ldots$; but these terms are independent, since the original observations were independent. More generally, knowing the payoffs of any finite set F of values gives no information about values not in F, which is what is meant by independence.

A PRETTY PROOF

In Chapter 19, John Halton found the answer to the natural "everyday problem" of the most economical way to lace a shoe. The following exposition, written by Michal Misiurewicz, gives a variation and extension of Halton's result.

LACING IRREGULAR SHOES

Michal Misiurewicz

Halton considered the problem of the most efficient (in terms of the length of the lace) lacing of a shoe. Let us recall the problem. We are given $2(n + 1)$ distinct points (the *eyelets*) $A_0, A_1, \ldots, A_n, B_0, B_1, \ldots, B_n$ in the plane. A *lacing* is a path $C_0 \to C_1 \to \cdots \to C_{2n+1}$, where $\{C_0, C_1, \ldots, C_{2n+1}\} = \{A_0, A_1, \ldots, A_n, B_0, B_1, \ldots, B_n\}$, $C_0 = A_0$, $C_{2n+1} = B_0$, and the points of the sets $\{A_0, A_1, \ldots, A_n\}$ and $\{B_0, B_1, \ldots, B_n\}$ alternate along the path. Its *length* is $|C_0C_1| + |C_1C_2| + \cdots + |C_{2n}C_{2n+1}|$. It is the length of the lace used, not including the slack pieces necessary to tie the lace. The problem is to find the lacing of minimal length.

Halton shows that the *standard lacing*[1] is the shortest one. The standard lacing is the lacing $A_0 \to B_1 \to A_2 \to B_3 \to \cdots \to A_{n-1} \to B_n \to A_n \to B_{n-1} \to \cdots \to A_3 \to B_2 \to A_1 \to B_0$ if n is odd, and $A_0 \to B_1 \to A_2 \to B_3 \to \cdots \to B_{n-1} \to A_n \to B_n \to A_{n-1} \to \cdots \to A_3 \to B_2 \to A_1 \to B_0$ if n is even (see Fig. 22.1).

The purpose of this section is to give a short proof of a considerably more general result. Halton assumes that the eyelets are arranged in two parallel rows with equal distances between consecutive eyelets. More precisely, there are constants $v, w > 0$ such that up to isometry, $A_i = (iv, w)$ and $B_i = (iv, 0)$ (see Fig. 22.2). This is a crude approximation of a real situation (see Fig. 22.3), where both "horizontal" and "vertical" distances between eyelets may vary. Even the symmetry with respect to a horizontal line can be absent, especially if the shoe is old.

---

[1]Halton calls this lacing *American style*. However, I do not recall anybody, be it in America or in Europe, having his or her shoes laced differently.

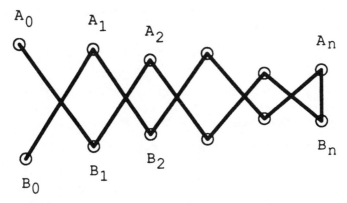

FIGURE 22.1. Standard lacing.

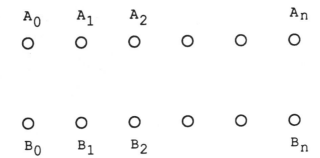

FIGURE 22.2. Halton's arrangement of eyelets.

FIGURE 22.3. A real shoe.

I want to propose a proof of minimality of the standard lacing under an assumption that seems to be much closer to the real situation:

(1)   For any k, l, the line through A_k and B_l separates the sets
$$\{A_i : i < k\} \cup \{B_j : j < l\} \text{ and } \{A_i : i > k\} \cup \{B_j : j > l\}.$$

In other words, if we view the "A" eyelets from any "B" eyelet (or the "B" eyelets from any "A" eyelet), they are in the natural order: $A_0, A_1, \ldots, A_n$ (or $B_0, B_1, \ldots, B_n$).

In fact, even this assumption is too strong. We need only the following:

(2)   For any $i < j$ and $k < l$, $|A_iB_k| + |A_jB_l| < |A_iB_l| + |A_jB_k|$.

To see that (1) implies (2), take $i < j$ and $k < l$, and assume that (1) holds. Then, as in Figure 22.4, the eyelets A_i and B_l lie on opposite sides of the line through A_j and B_k, and the eyelets A_j and B_k lie on opposite sides of the line through A_i and B_l. Therefore, the segment A_iB_l intersects the segment A_jB_k. If D is the point of intersection, then by the triangle inequality we get

$$|A_iB_k| + |A_jB_l| < |A_iD| + |DB_k| + |A_jD| + |DB_l| = |A_iB_l| + |A_jB_k|;$$

so (2) holds.

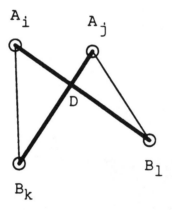

FIGURE 22.4. (1) implies (2).

One can consider an even more realistic setup, where the eyelets are points of three-dimensional space. However, a lacing should closely follow the surface of the shoe. Therefore, we can assume that the eyelets are points of the surface of the shoe, with distances measured along this surface. The reader can make precise measurements to verify whether in this setup his or her shoes satisfy assumption (2). But usually the piece of the shoe surface involved in lacing has a very small curvature. This means that we do not make a big error if we view this surface as a "bent" piece of the plane. Smoothing it out leads us to our initial model. Now a quick glance is enough to verify whether assumption (1) is satisfied.

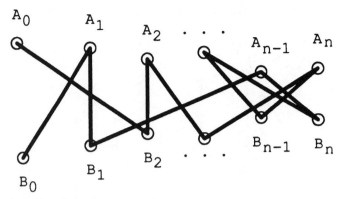

FIGURE 22.5. An arbitrary lacing.

It is time to prove the main result of this section.

Theorem. *Assume that the set of eyelets* $\{A_0, A_1, \ldots, A_n, B_0, \ldots, B_n\}$ *satisfies* (2). *Then the standard lacing is shorter than any other lacing.*

The idea of the proof is rather simple. Referring again to Figure 22.4, let us define a "move" to consist in replacing the pair of segments A_iB_l and A_jB_k by A_iB_k and A_jB_l. The only problem is that the result of such a move may not be a lacing. For instance, if we start with the lacing shown in Figure 22.5 and make a move that replaces segments A_0B_2 and $A_{n-1}B_1$ by A_0B_1 and $A_{n-1}B_2$, then what we get (Fig. 22.6) is not a lacing. Therefore, we will introduce objects that are more general than lacings, namely systems of arrows. In this larger class, all moves will be legitimate, and after a finite number of moves, we arrive at the system corresponding to the standard lacing.

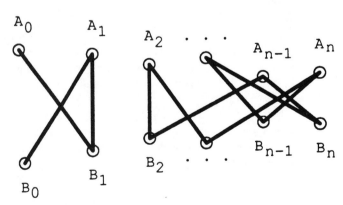

FIGURE 22.6. Result of a move.

These ideas lead to the following formal proof.

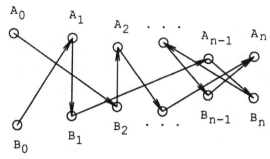

FIGURE 22.7. A system of arrows obtained from the lacing in Figure 22.5.

Proof of Theorem: Let $L = C_0 \to C_1 \to \cdots \to C_{2n+1}$ be a lacing (see Fig. 22.5). There is some k such that $C_k = A_n$. We draw arrows from C_i to C_{i+1} for $i = 0, 1, \ldots, k-1$ and from C_{i+1} to C_i for $i = k, k+1, \ldots 2n$ (see Fig. 22.7). Note that this system of arrows satisfies the following properties.

(3) Each arrow begins at an "A" eyelet and ends at a "B" eyelet or vice versa. All eyelets except A_0, B_0, and A_n have one arrow coming in and one going out. Eyelets A_0 and B_0 have one arrow going out and none coming in. Eyelet A_n has two arrows coming in and none going out.

We define the length of a system of arrows as the sum of their lengths. Although not every system of arrows comes from a lacing, if it does, then its length is equal to the length of the corresponding lacing.

Let us call the system of arrows obtained from the standard lacing the *standard system*. We show that it has smaller length than any other system of arrows satisfying (3). Indeed, if this is not the case, then there exists a system of arrows $\mathcal{A}$ satisfying (3) that has minimal length and is not standard. If the arrows $A_k \to B_{k+1}$ and $B_k \to A_{k+1}$ for $k = 0, \ldots,$ $n-1$ are in $\mathcal{A}$, then by (3), $\mathcal{A}$ consists of these arrows and the arrow $B_n \to A_n$. This means that the system $\mathcal{A}$ is standard, a contradiction. Therefore, we can take the smallest k such that at least one of the arrows $A_k \to B_{k+1}$, $B_{k+1} \to A_{k+1}$ is not in $\mathcal{A}$. We may assume that this is $B_k \to A_{k+1}$. Thus, we must have in $\mathcal{A}$ arrows $B_k \to A_j$ and $B_l \to A_{k+1}$ for some $j \neq k$ $+ 1$ and $l \neq k$. Since any arrow ending at A_j with $j \leq k$ begins at B_{j-1} and any arrow beginning at B_l with $l < k$ ends at A_{l+1}, we have $j > k+1$ and $l > k$. If we now make the move replacing the arrows $B_k \to A_j$ and $B_l \to A_{k+1}$ by the arrows $B_k \to A_{k+1}$ and $B_l \to A_j$, the new system will still satisfy (3). However, by (2) (with $i = k-1$), its length will be smaller than the length of $\mathcal{A}$, a contradiction. This completes the proof.

The Sun, the Moon, and Mathematics

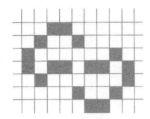

When I heard the learn'd astronomer,

When the proofs, the figures, were ranged in columns before me,

When I was shown the charts and diagrams, to add, divide, and measure them,

When I sitting heard the astronomer where he lectured with much applause in the lecture-room,

How soon unaccountable I became tired and sick,

Till rising and gliding out I wander'd off by myself,

In the mystical moist night-air, and from time to time,

Look'd up in perfect silence at the stars.

—Walt Whitman

I hesitate to pick a fight with Walt Whitman, who's dead and can't defend himself, but I have to wonder why he was so turned off by a few proofs and figures. The learn'd astronomer, after all, was just doing his job, trying to find out how things work. I have a feeling, though, that this poem is comforting to many people, who become dismayed, perhaps even becoming "tired and sick," when confronted by science. To hear Walt tell it, the universe somehow became less "mystical" when Copernicus made his discoveries about the solar system. Well, I'm sorry, Mr. Whitman, but I don't believe knowing the laws of celestial mechanics need affect a person's ability to react on an emotional level to the beauty of the night sky. Indeed, if anything, it's the other way round. That other great poet had it right, it seems to me: Truth *is* beauty (and beauty truth)—and a good thing it is for all of us, poets and astronomers alike.

But thanks anyway, Walt, for a poem about the night sky, which will be the text for this mini-sermon. The sermon itself, besides expositing some elementary astronomical facts, is intended to illustrate a few quasi-philosophical points which have intrigued me over the years. The first is the fact that mathematics is all around us. Theorems abound, and to find them, all we need to do is keep our eyes and minds open. The second point is about avoiding numbers. There is a tendency not just among mathematicians but among people in general to try to quantify everything, but, as I have illustrated in the past, sometimes introducing numbers hinders rather than helps understanding. Therefore, and the Whitmanites should be grateful, I will not "add, divide, and measure." Finally, perhaps I should have put a question mark after the word "mathematics" in my title. The reader will have to decide whether the discussions to follow are mathematics or just "common sense."

THE MOON

All of us have "from time to time look'd up" and observed a crescent moon. A contemporary poet might compare its shape to a clipped fingernail, but in less poetic language it can be described as a region bounded by two curves, which we will refer to as its *outer* and *inner* boundaries. The mathematical (?) problem is then to describe these curves. Are they arcs of circles, of ellipses, or what? After a few false starts, trying to introduce coordinates and the like, I drew the standard picture that many of us no doubt saw when we were children (I hope it's not one of those "diagrams" that old Walt found so upsetting). We are looking directly at the moon from earth, perhaps through a telescope. The sun is behind and to the lower left of the plane of the picture. The outer and inner boundaries are denoted by B_O and B_I.

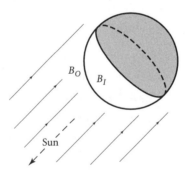

The story is now a simple one. We consider two hemispheres on the moon. The first is the *visible* hemisphere, the one that can be seen from the earth. We are looking directly at it in the picture. The second is the *bright* hemisphere, the one that is illuminated by the sun. The intersection of these two hemispheres, called a lune, is what we see.

The answer to our question is now clear. The outer boundary is a semicircle whose diameter is the diameter of the moon. The inner boundary is a "semiellipse" whose major axis is the diameter of the moon. This is because it is the projection of the "semiequator" of the moon, as shown in the diagram. What is the minor axis, b, of B_I? For this we will need another diagram.

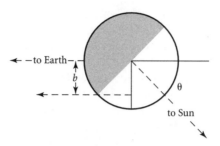

The plane of the diagram is the plane determined by Earth, Moon, and Sun. If r is the radius of the moon and θ is the angle between the rays Earth–Moon and Moon–Sun, then one sees that the minor axis, b, is given by $b = r(\cos\theta)$.

An interesting special case is that of the half-moon, when θ is a right angle. In that case let α be the angle between the rays Earth–Moon and Earth–Sun. The great Greek astronomer Eratosthenes used this to calculate the ratio of the distance of the earth from the moon and from the sun. Namely, choosing a time when both the moon and the sun were visible, he measured the angle α between them (as seen from the earth). Then the ratio of distances to sun and moon is simply $\cos\alpha$. This remarkable man was also able to calculate the diameter of the earth from astronomical observation. His estimate turned out to agree with our current estimate to within fifty miles!

We remark that the answer to our question provides an example of an ellipse that "occurs in nature." Are there others, I wonder, aside from the orbits of the planets?

THE SUN

Since I said I would be talking about the night sky, you may wonder how the sun gets into the picture. At this point we depart from most people's everyday experience, that is, unless you happen to be a person who lives above the Arctic Circle, in which case there will be times in summer when the sun shines twenty-four hours a day. Indeed, I witnessed this myself on a recent Norwegian cruise and found the experience quite awe-inspiring (I'm sure Walt Whitman would have reacted similarly). But it also led me to some questions, rather simple ones to be sure, but nevertheless mathematical. Or were they? First of all, I said to myself, below the Arctic Circle the sun rises in the east and sets in the west, but what happens when the sun neither rises nor sets? Specifically, at which point of the compass will the sun be when it's at its lowest point, that is, closest to the horizon? One question led to another, and I then wondered what exactly the shape of the sun's "apparent orbit" was. This question, of course, needs to be made precise. Let's think of it this way. You stand in a fixed place, equipped with a telescope, and probably a pair of strong sunglasses. At each instant you keep your telescope pointed at the sun, so in the course of the day the telescope sweeps out some sort of a cone, returning to its original position after twenty-four hours. We shall refer to this as the *sun-cone*. What sort of a cone is it? Circular? Elliptic? If circular, where is its center? What is its radius?

The figure below gives the answers. We are looking at the projection of the earth in the plane determined by the sun and the earth's axis (a point and a line determine a plane). The sun is off at infinity in the horizontal direction. The learn'd astronomer, represented by $\overline{\overset{\circ}{\Upsilon}}$ (a concession to Walt, to show that astronomers really are human), is standing at latitude ɸ, and it is "midnight," that is, the time when the sun is closest to the horizon, and the telescope, →, is pointing at the sun, which is due north. The normal to the sphere at the point where the astronomer is standing gives the direction to the zenith, as shown, and the angle between the telescope and the zenith is α. On the same diagram at noon, twelve hours later, the sun is overhead and slightly to the south, and the angle between the telescope and the zenith is β. One could try to sketch the positions of the astronomer at other times of the day, but this would only serve to clutter the picture.

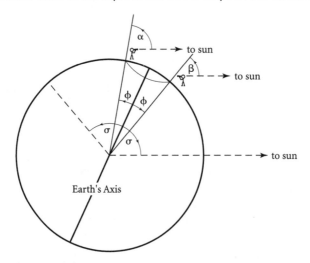

The figure above shows what is really happening in a coordinate system in which the sun is fixed. Our problem is now to translate this to a system in which the pre-Copernican astronomer is fixed and the sun does the moving. Clearly, the only thing that matters is the angle between the position of the sun and the normal to the earth at the point where the observer is standing. Let us assume that the observer stays facing north, like the two figures at the top of our diagram. It is also clear that the sun-cone is completely determined by two parameters: the latitude, or more conveniently the colatitude, ɸ, where the astronomer is standing; and second, the angle of "tilt" of the earth, that is, the angle σ that the earth's axis makes with the line from earth to sun. It is now easy to see that the angle α of the figure is σ+ɸ and the angle β is σ−ɸ, so the generator of the cone makes the angle $(\alpha+\beta)/2 = \sigma$ with the earth's axis. The picture now tells the story. In the fixed-sun description the ɸ-cone along with the observer rotates about the earth's axis. In the fixed-observer description it is the σ-cone that does the rotating about the observer, as indicated by the dashed line in the diagram.

To summarize, the sun-cone as seen by the observer/astronomer is seen as a cone of revolution with generating angle the tilt angle σ, and the central ray of the cone points in the direction of the earth's axis, so its angle relative to the observer is ɸ, the observer's colatitude; or equivalently, the central ray of the cone makes an angle with the horizon that is equal to the observer's latitude. Thus, the "radius" of the cone depends only on the earth's tilt, hence on the "season," that is, the position of the earth relative to the sun; but it is independent of the observer's position (latitude), while the center depends only on the observer's position and is independent of the season.

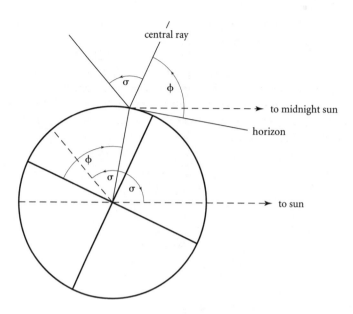

Let's dress this up a little and try to make it look more mathematical.

Theorem. *The perceived orbit of the sun is a cone of revolution. The center of the cone is given by a ray making an angle ϕ° with the ray from the observer to the horizon, where ϕ° is the latitude of the observer. The generating angle of the cone is σ, where σ is the tilt angle of the earth toward the sun.*

Please note that the midnight sun was merely the motivation for the inquiry. The theorem is true generally whether or not the sun goes below the horizon.

MATHEMATICS

Or was it? We seem to have answered all of our questions, and the questions and answers sounded mathematical, with ellipses, circles, angles, etc.; but what did we use besides a few pictures and a bit of common sense? Clearly, we have not presented anything resembling formal proof of our assertions. Presumably, it would have been possible to do this by carefully modeling the various astronomical objects as subsets of Euclidean 3-space, but it seems to me that such a project would not be worth the effort. In fact, a reader who was sufficiently interested to go through such a proof would probably have to translate the symbols back into the intuitive objects of our argument in order to "see" the answer.

Note also the complete absence of numbers in our results. This may not be quite so obvious. One is likely, for instance, to think of latitude as a number, but this is not necessary and is in a way misleading. Latitude is an angle (the latitude of a point on the earth is the complement of the angle between the normal to the earth at that point and the earth's axis), and an angle is not a number. An angle is an ordered pair of rays emanating from a common point. Likewise, the major and minor axes of an ellipse are not numbers.

They are segments. As I have remarked before, the concept of number never occurs in Euclid's elements, and it is not needed here.

And finally, to come back to our starting point; yes, Walt Whitman, it is good to be reminded of the awe one experiences in gazing in perfect silence at the night sky. But there are also some of us who get a feeling of satisfaction from such seemingly useless activities as learning the exact shape of a crescent moon. The point is that the sources of wonder are all about us, but they come to us in many different forms, and we are fortunate if we can be open to all of them.

In Praise of Numberlessness

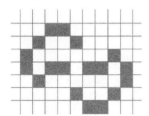

AVOIDING NUMBERS

One of the more frustrating things about being a mathematician is that almost no one other than professional scientists has the faintest idea what the subject is about or what it is we do. Ask a person chosen at random what the central concept of mathematics is and the answer will probably be *numbers*. Now, I don't know myself what mathematics is about (more on this later), but I'm quite sure that the central concept is not numbers. Still, there is a central concept. It is *proof*. The examples to follow are intended to illustrate how sometimes the reflex action of trying to use numbers makes proofs more difficult.

COUNTING BEANS

At the most naive level, consider the problem of deciding which of two jars contains more beans. Easy, you say, just count the number of beans in each jar. True, but a person who didn't know how to count might find the quicker solution. Remove pairs of beans simultaneously from each jar, and stop when one of them is empty.

This fable illustrates (1) the advantage of "parallel computation" and (2) at a very primitive level, the notion of cardinality of sets via one-to-one correspondences.

MIXING WINE

A bit more sophisticated is this well-known problem, in which the same idea is illustrated. You are given a glass of red and a glass of white wine and are told to take a spoonful of the red, add it to the white, and mix completely. Then take a spoonful of the mixture and return it to the red. *Question:* Is there more red in the white or white in the red? (*Warning:* this problem may have the effect of alienating people, who may argue vehemently for wrong answers like, "You added pure red to the white but a mixture to the red so there must be more red in the white.")

Now of course one can work it out quantitatively. Let R and W be the given volumes of red and white wine (measured in some units) and let S be the volume of the spoon. Then it is not hard to get the formula for the amounts in question in terms of R, W, and S. The point is that this is all unnecessary, and, further, the condition of complete mixing is a red herring. Here is an alternative scenario. You have a full glass of wine (of either color). Throw some of it in the ocean, wait 5 minutes, and then refill the glass from the now slightly polluted ocean. *Question:* Is there more wine in the ocean or (pure) ocean in the wine?

Answer: The two amounts are equal. *Reason:* At the end of the experiment the ocean in the wine glass has *replaced* the wine that was originally there. *Exercise:* Where is the missing wine?

The significance of this example is that it not only avoids numbers but, one can argue, it avoids mathematics of any sort, using nothing but *common sense.* I will return to this matter in the next section.

QUALITATIVE GEOMETRY

The preceding questions and those to follow are *qualitative,* as opposed to quantitative, problems. A qualitative question deals with concepts like *more, fewer, greater, equal, less, bigger, smaller,* whereas a quantitative problem involves attaching numbers to objects. The example par excellence of a qualitative theory, and I was surprised to realize this, is the geometry of Euclid. This is not to be confused with what we now refer to as Euclidean geometry with the "Euclidean metric." There is no metric, no concept of distance, no notion of the *size* of an angle in Euclid's *Elements.* It is true that Euclid adds, subtracts, and compares objects, but these objects are the segments and angles themselves, not their lengths or sizes. Here, for example, is the triangle inequality, the celebrated Proposition XX.

> *In any triangle two sides taken together are greater than the remaining side.*

Here is Euclid's proof:

Given triangle *ABC*, extend side *BC* to *D* so that *DC* equals *AC* (here and elsewhere for "equals" read "is congruent to").

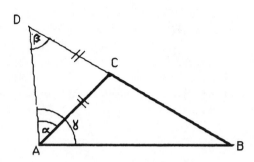

Then, since triangle ADC is isosceles, angles α and β are equal (Proposition V). Further, α is less than (read "contained in") γ. But in triangle ABD, the side opposite γ is $AC + CB$, and the side opposite β is AB. So by (previously proved) Proposition XIX, $DC + CB = AC + CB$ is greater than AB.

Qualitative all the way! But let us continue to backtrack. Proposition XIX is the immediate converse of

Proposition XVIII. *In any triangle, the greater side subtends the greater angle.*

Referring to the figure below, we suppose that AB is longer than BC.

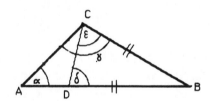

Choose D on AB such that BD equals BC. Then CBD is isosceles. So angles ϵ and δ are equal (Prop V). But ϵ is contained in, hence less than, γ, and δ, an exterior angle of triangle ADC, is greater than α.

This, in turn, used

Proposition XVI. *The exterior angle of a triangle is greater than either of the interior opposite angles.*

Proof.

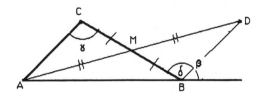

To show that β is greater than γ, connect A to the midpoint M of BC and extend the line AM to D so that MD equals AM. Then triangles AMC and DMB are congruent (side–angle–side), so γ equals δ, which is contained in, hence less than, β. ■

Remark. We know, and so did Euclid, the much stronger fact that the exterior angle is the sum of the opposite interior angles (Proposition XXXII). Why then does Euclid settle for this much weaker result? The reason is surely that he wants to go as far as possible without having to make use of the parallel postulate. In today's language we would say that the three propositions proved above are the theorems of absolute geometry, holding in elliptic and hyperbolic as well as Euclidean geometry. Once again we see an example of the astonishing sophistication of the mathematics of 2,000 years ago.

STACKING RECTANGLES

Two congruent rectangles are stacked one on top of the other as shown in the figure below on the left. *Question:* Does the upper rectangle cover more or less than half of the lower one?

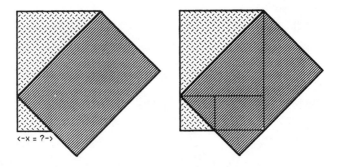

The first impulse is probably to bring on the numbers, i.e., calculate the areas of the two uncovered triangles. One can do this, of course, but even after arriving at the somewhat messy expression, it is not at all obvious which way the inequality should go, *whereas* a little thought suggests drawing the two lines shown in the figure on the right and observing the two pairs of congruent triangles. Indeed, one does not even need to have the concept of area to present the problem. We need only agree the one region is bigger than another if the second is "piecewise congruent" to a subset of the first.

WHAT IS MATHEMATICS ANYWAY?

After much rumination I've reached the conclusion that there's no such thing as mathematics. Let me elucidate. Consider again the wine–water problem. If one "deconstructs" it properly, all the mathematics disappears.

Here is another well-known question that doesn't seem to be mathematics at all . . . or is it?

Doing It with Mirrors

Why is it, people have asked, that mirrors reverse right and left but not top and bottom?

This hardly seems like a question in mathematics, but if not, what kind of a question is it? Physics? Optics? There is a fairly substantial literature on this question, but I have yet to see a completely satisfactory discussion. Here's how I look at it.

1. The mirror is irrelevant. The phenomenon occurs in other quite different contexts. Write some words on a thin sheet of paper and then turn it away from you and hold it up to the light. The words are still there, but the writing runs backward, not upside down. Is this a strange property of paper? Of course not. The phenomenon was produced by you, the experimenter, when in turning the paper toward the light you chose to turn it about a vertical rather than a horizontal axis.

2. My local barbershop has a sign painted on its window that viewed from the street looks like this,

but as I look at it from inside while the barber is at work, it looks like this,

but not like this.

How come? Again it was I, the "experimenter," who caused this to happen, but in this case I moved myself rather than the window. In looking at the window from the inside I was facing in the opposite direction from which I looked at it from the street, and I achieved this by turning myself about a vertical axis. If I had wanted to and was sufficiently well coordinated, I could have faced the window from inside the barber shop by standing on my head, in which case the writing would have appeared to be like the last figure above.

3. The mirror phenomenon is felt to be paradoxical because the mirror seems to reverse sides but not top and bottom. But this is not true. My mirror is on the north wall of my bedroom, and my window is on the west wall, looking out at San Francisco Bay. When I look in the mirror, the reflected window is also on the west wall of the reflected room. Clearly, the mirror does not reverse east and west. Why then are we confused?

4. Nevertheless, you may say, why is it that the guy I am looking at in the mirror seems to be wearing his watch on his right wrist (rather than his left ankle).

The point is that these paradoxes are of our own making. *Right* and *left,* unlike east and west or top and bottom, are in the eye of the beholder. So are *backward* and *upside down.* All of the italicized words are defined *relative to the observer,* whereas east, west, north, south, up, down are objective. The sun always rises in the east, no matter where I happen to be, but it may rise on my right or left or in front or in back depending on my position. The fallacy is in mixing relative with absolute concepts, the "subjective" left–right with "objective" east–west. When I say that my reflection wears a watch on his right hand rather than his left ankle, it is because if I were to replace "him" in the mirror, I would do so by making an about-face, keeping my feet on the ground, rather than going into a handstand. The fact that we humans, along with most other animals, are bilaterally symmetric makes us feel that the only natural way to "turn around" is about a vertical rather than a horizontal axis, but it doesn't have to be that way.

But let us get back to the issue at hand. In what way can the arguments above be thought of as mathematical? I claim there is some similarity both in form and content. As to form, we have insisted, as in mathematics, on precise definitions of the words we use. As to content, the key words, left and right, are the everyday expression of the notion of orientation, which is pervasive in much of mathematics.

Here is a last example, which is even more remote from what one would normally consider mathematics. I recall reading in a child's science book the fairly obvious optical fact that the images of objects on the retina of the eye are upside down. The book went on to say that the brain by some remarkable unexplained trick manages to turn them right-side up. This is another example of the subjective–objective fallacy. To whom does the image on the retina appear upside down? To someone peering into my eye from the outside. To say that it appears upside down to me implies the absurd idea that I am looking at my own retina.

The point of these examples is to illustrate the gray area between what is and is not mathematics, but, more importantly, to show how we get tricked and trapped by words if we fail to analyze them rigorously enough. It is my belief that we are victims of a similar word trap when we try to define mathematics.

Looking for a definition of mathematics is chasing a will-o'-the-wisp. We don't even have a definition of, for example, the word "green." My wife insists that this green tie I'm wearing is actually blue, and there is no way either of us can prove we are right. On a more highbrow level, I recall that many years ago there was an ongoing exchange of letters in the *New York Times* Book Review section on the question "What is History?" Scholar A said history is simply everything that happens, but scholar B insisted that, no, history is only those things that happen that make other things happen, or words to that effect— and the debate went on. I think these things illustrate what I call the linguistic fallacy. We invent words like "history," "mathematics," "green," to communicate with each other. But then we think that because the words exist they must correspond to some actual outside entity, that the concept of History or Mathematics has some sort of existence of its own, perhaps in the mind of God, and so we seek to discover what these things *really* are, forgetting the fact that it was we, not He/She, who invented them in the first place.

What is mathematics? My answer is that there is no such *thing*. For some things, e.g., up, down, left, right, it is crucial to have definitions. For others, like mathematics, searching for a definition becomes mere wordplay. Let's not waste time on it.

AND A FINAL PERSPECTIVE

This from a granddaughter now in second grade who is somewhat ambivalent on the subject of mathematics. As she puts it, she likes carrying, but she can't stand borrowing— which probably says it all. Over and out.

A Curious
Nim-Type Game

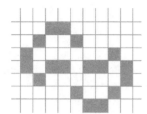

A set of mn objects is laid out in an $m \times n$ rectangular array. We denote by (i,j) the object in row i, column j. The first player, P_I, selects an object (i_1,j_1) and then removes all objects (i,j) such that $i \geqq i_1$ and $j \geqq j_1$. In other words, if i increases upward and j increases from left to right, then P_I removes a northeast "quadrant." Player two, P_{II}, now picks (i_2,j_2) from among the remaining objects and removes all (i,j) such that $i \geq i_2$, $j \geq j_2$. The play then reverts to P_I and continues in the same way until all objects have been removed. The player making the last move loses. Thus the object of the game is to make your opponent pick up $(1,1)$.

There are some trivial special cases of the game.

Case A: P_I wins the $2 \times n$ ($m \times 2$) game by selecting $(2,n)$ $((m,2))$. Then, whatever P_{II} does, P_I moves so as to leave a "position" in which there is one more object in row (column) 1 than in row (column) 2. The reader will easily see that this is always possible and winning.

Case B: P_I wins the $m \times m$ game by selecting $(2,2)$. From then on he "symmetrizes." Whenever P_{II} chooses $(1,j)$, he chooses $(j,1)$, etc. Again it is easily seen that this wins.

The above are the only two cases in which general winning strategies are known. What makes the game interesting, however, is the following:

Theorem. *For all m and n, the game is a win for P_I.*

The proof of this fact is typical of something that occurs quite often in game theory in that it is completely nonconstructive. Although it establishes the existence of a winning strategy for P_I, it is of absolutely no use in finding such a strategy. Here is the argument. There are two possibilities.

Case 1: P_I has a winning strategy in which his first move is to select (m,n).

Case 2: If P_I selects (m,n), he loses. Then there must be a response (i_2,j_2) by P_{II} that wins for P_{II}. This means that the position of the game after P_{II}'s move is a loss for the player who must then move, in this case P_I. The point, however, is that P_I could have handed this position to P_{II} if he had himself chosen (i_2,j_2). Hence (i_2,j_2) is a winning first move for P_I, and the assertion is proved.

The above argument is reminiscent of the well-known one that asserts that games like tic-tac-toe cannot be a win for player II (it applies, for instance, to the unsolved go Moku, which is tic-tac-toe, 5 in a row on a 19×19 board.) It goes like this. Suppose the game were a win for P_{II}. Then let P_I make any first move, and then pretending in his mind that he did not make it, from then on behave in accordance with the alleged winning strategy for P_{II}. This will always be possible unless at some point this strategy requires him to move onto the square he chose initially. In that case, he again makes an arbitrary move. We see then that the alleged winning strategy for P_{II} is also available to P_I, but by definition of winning there cannot be winning strategies for both players. Contradiction. (Note, however, that the tic-tac-toe proof as given here is a proof by contradiction, while the proof of our theorem is direct, which points up the interesting fact that a nonconstructive proof is not necessarily a proof by contradiction.)

It may be of interest to observe that both Nim and this game (Gnim? Gnome?) are special cases of the following general class of games: Let S be any partially ordered set. A player moves by choosing some element of S and removing all elements greater than or equal to it, and the player moving last loses. Nim corresponds to the special case where S is the *sum* (disjoint union) of a finite number of totally ordered sets. Gnim is the case where S is the product of two totally ordered sets. The argument showing that Gnim is a win for P_I applies to any set S that has a *largest* element, thus, for example, the product of any number of totally ordered sets, but of course not to Nim.

The above is essentially all the theoretical information I have about this game (I shall give one further special result at the end.) However, with the aid of a computer some rather intriguing empirical results have been obtained. The $3 \times n$ game has been completely analyzed for $n \leq 100$, and in all cases there is only one winning first move for P_I. This is also the case for 4×5 and 4×6 and, of course, Cases A and B discussed above.

QUESTION
Is the winning first move unique for all m and n? (See note.)

The diagrams below give the winning first move in the $3 \times n$ game for $2 \leqq n \leqq 12$.

```
X          X        XX       XXX       XXX       XXXX      XXXX
XX         X        XX       XXXXX     XXX       XXXXXXX   XXXX
XX         XXX      XXXX     XXXXX     XXXXXX    XXXXXXX   XXXXXXXX
2          3        4        5         6         7         8


XXXXX               XXXX              XXXXXX             XXXXXXX
XXXXXXXXX           XXXX              XXXXXX             XXXXXXXXXXX
XXXXXXXXX           XXXXXXXXXX        XXXXXXXXXXX        XXXXXXXXXXX
9                   10                11                 12
```

There are, of course, two types of moves depending on whether one takes the initial "bite" from the top row only or from the top two rows. It turns out that roughly 58 percent of the moves are of this second type, as for the cases $n = 3, 4, 6, 8, 10, 11$, while 42 percent are of the first type, e.g., $n = 2, 5, 7, 9, 12$. In general, the length of the bite appears to increase with n. In fact, for all $n \leq 170$, there is only one counterexample. For $n = 87$ the bite size is 37, while for $n = 88$ the bite size is 36 (both of these are two-row bites). Phenomena like this lead one to believe that a simple formula for the winning strategy might be quite hard to come by.

We close with a conjecture: *It is never optimal to select (m,n) on the first move except when $m = 2$ or $n = 2$.* We shall prove this for the case $m = 3$ as a further illustration of the type of nonconstructive argument one uses. This one requires an argument by contradiction.

First observe that in the 3×3 game $(3,3)$ is losing (since $(2,2)$ wins). Now assume that $(3,n)$ is losing for all $3 \times n$ games up to $\bar{n}$ and consider the case $3 \times (\bar{n} + 1)$. The argument is best given with pictures:

suppose | Fig. 1 | X is a winning position
X

$\leftarrow \bar{n} \rightarrow$

then | Fig. 2 | X X must be losing,
X X

$\leftarrow \bar{n} - 1 \rightarrow$

so there must be a way of going from Fig. 2 to a winning position. Now clearly no choice (i,j), $j \leq \bar{n}$, gives a winning position, for this would leave a position that P_{II} could have presented to P_{I}, contradicting the assumption that $(3, \bar{n} + 1)$ was winning for P_{I}. The only possible choices are therefore either $(\bar{n}, 1)$ or $(\bar{n}, 2)$, but $(\bar{n}, 1)$ leaves

X
X '

$\leftarrow \bar{n} - 1 \rightarrow$

which loses by the induction hypothesis, and $(\bar{n}, 2)$ leaves

$$\boxed{}\begin{matrix}\text{X}\\\text{XX}\end{matrix} \; ,$$
$$\leftarrow \bar{n} - 1 \rightarrow$$

after which P_{II} can play $(3,1)$, leaving

$$\boxed{}\begin{matrix}\text{X}\\\text{XX}\end{matrix} \; ,$$

which loses for the $2 \times (n + 1)$ game of Case A, so the proof is complete.

One can prove any number of special results of this sort by similar arguments. For example, for $n > 4$, it is never winning to choose $(2, n - 1)$. For $n > 5$ it is never winning to choose $(3, n - 1)$, and presumably some general inequalities exist showing that for large rectangles the bites cannot be too small. I expect that the problem of finding explicit winning strategies may be hopeless.

Finally, let me mention some generalizations. The first is to allow either m or n or both to be infinite. However, these games are rather trivial, because (a) $1 \times \infty$ is a win for P_I (trivial), (b) $2 \times \infty$ is a win for P_{II} (a nice exercise for the reader), and (c) $m \times \infty$, $2 < m \leq \infty$, is a win for P_I, as he can choose $(2,1)$, leaving P_{II} with $2 \times \infty$. Of more interest are higher-dimensional games, e.g., $m \times n \times r$ in 3-space. Of course, any such finite game is solved in a finite amount of time by, at worst, enumerating all possible strategies. The real challenge, it seems to me, is games like $3 \times 3 \times \infty$ or even $\infty \times \infty \times \infty$. ($2 \times m \times \infty$ is a win for P_I. Why?) These belong to an interesting class of games with the property that although every play of the game terminates after a finite number of moves, there is no upper bound on the possible lengths of a play (as there is for example in chess). In particular, I don't know of any way to program a computer to find out say, if $3 \times 3 \times \infty$ is a win for P_I or P_{II}.

Note: A description of Gnim appeared in the column by Martin Gardner in *Scientific American*, pp. 110–111, January 1973. In response to the article, K. Thompson of Bell Laboratories and M. Beeler at M.I.T. discovered by using computers that there exist games with more than one winning first move. The smallest known example is 8×10. Further, it was learned that a numerical game isomorphic to this one was described by F. Schuh in an article entitled "The Game of Devisors" in *Nieuw Tijdschrift voor Wiskunde*, 39, pp. 299–304 (1952).

The Jeep Once More
or Jeeper by
the Dozen

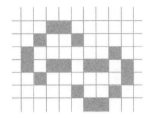

1. INTRODUCTION

In 1947, N. J. Fine solved the now famous problem of the jeep [1]. The problem concerns a jeep that can carry enough fuel to travel a distance d, but is required to cross a desert whose distance is greater than d (for example $2d$). It is to do this by carrying fuel from its home base and establishing fuel depots at various points along its route so that it can refuel as it moves further out, and it must cross the desert on the least possible amount of fuel.

Our purpose here is first to give a very short derivation of the solution of the jeep problem that makes use of a theorem (also famous) of Banach [2]. Second, we consider the situation in which it is required to send a jeep across the desert every day, say, for a week. Of course, having found the best procedure for a single jeep, one could simply repeat this seven times. We shall show, however, that there is a more economical procedure for the case of several jeeps, and in general that the more jeeps one sends across, the lower fuel consumption per jeep. This phenomenon is an example of what economists refer to as "increasing returns to scale," a subject of some economic interest. (It also accounts for the alternative title of this article, for which I apologize herewith.)

2. THE PROBLEM

To formalize the problem, let us assume that the jeep starts from the origin and moves along the positive x-axis. We choose for the unit of fuel the maximum amount that the jeep can carry and refer to this unit as a *load*. The unit of distance is then chosen as the distance the jeep can travel on one load.

Figure A2.1 is a schematic representation of a typical jeep's journey.

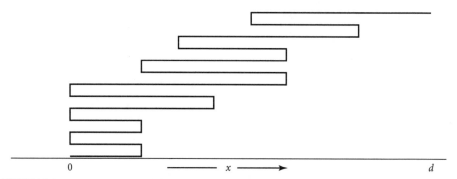

FIGURE A2.1.

The wiggly path represents the jeep's travel. Of course, actually the path lies entirely on the x-axis. It has been stretched vertically simply to make it visible. Because of our choice of units, the length of this path is precisely equal to the amount of fuel consumed. In the figure, the jeep reaches a point at distance d from the origin. The original jeep problem asks for the minimum fuel consumption (hence path length) that will allow the jeep to reach this point. It is somewhat more convenient to turn the problem around (see, e.g., [3]) and obtain a formula for the function $d(f)$ giving the farthest point that the jeep can reach on f loads of fuel. Our first aim is to prove the formula

(1)
$$d(f) = 1 + \frac{1}{3} + \frac{1}{5} + \cdots + \frac{1}{2f-1}.$$

The key idea in our solution is to use a formula of Banach for the path length of a curve in one-dimensional space. (Path length in one-space is usually referred to as *total variation*; we prefer the geometric terminology as being more suggestive in the present context.) To utilize Banach's formula we define, for each point x on the interval $[0,d]$, the *valance* $n(x)$ as the number of times during its journey that the jeep is at point x. Figure A2.2 gives the graph of the valance $n(x)$ corresponding to the jeep's journey plotted in Figure A2.1. Strictly speaking, if one allowed very general paths, $n(x)$ might be infinite for some points. This would not affect Banach's formula, which states that

$$\text{total path length} = \int_0^d n(x)\,dx.$$

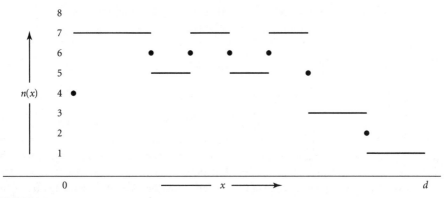

FIGURE A2.2.

Of course, Banach was not concerned with jeeps. He considered a continuous real-valued mapping $x = \phi(t)$, and $n(x)$ was simply the number of points that mapped onto the point x, or the cardinality of $\phi^{-1}(x)$. We can also think of the problem this way if we take the wiggly line of Figure A2.1 as the graph of the jeep's position plotted, say, as a function of the time.

For "reasonable" paths, Banach's formula is obvious. A reasonable path for a jeep has a finite number of points at which it reverses direction (it would be a remarkable jeep indeed that could execute an unreasonable path). To prove the formula, partition $[0,d]$ into sets $X_1, X_2, \ldots$, where

$$X_k = \{x \mid n(x) = k\}.$$

Because of reasonableness, there are only finitely many nonempty X_k, and each of them is a union of disjoint intervals. (Some of them may consist of single points. In the case of the jeep's tour, this is the case whenever k is even. Do you see why?) Exactly k intervals of the jeep's path lie over each interval of X_k. Therefore,

$$\text{total path length} = \sum_{k=n}^{\infty} k \, (\text{length of } X_k).$$

But the term on the right is precisely the definition of $\int_0^d n(x)\,dx$. (Notice that this is the definition of the integral in the sense of Lebesgue rather than Riemann! Of course for continuous functions and in particular for reasonable functions, the two concepts are equivalent.) We remark that Banach's formula holds for unreasonable as well as reasonable paths (though the general proof is fairly involved), and hence our solution of the jeep problem holds for both unreasonable and reasonable jeeps.

3. THE SOLUTION FOR ONE JEEP

We return now to the problem of computing $d(f)$ and assume for the moment that f is an integer. We wish to determine how far the jeep can travel on f loads of fuel. For any jeep's tour, we define the sequence of points $x_0, x_1, \ldots, x_f$ on the interval $[0,d]$, where

$x_f = 0$, $x_0 = d$, and in general, x_k is the point such that the total path length (hence fuel consumption) to the right of x_k is exactly k units. Clearly, the points x_k form a strictly decreasing sequence, and there is exactly one unit of path length between x_{k+1} and x_k. The basic observation we need is the following:

Lemma 1. *If $x < x_k$, then*

(2) $$n(x) \geq 2k + 1.$$

Proof. Since x is to the left of x_k, the jeep must consume more than k loads of fuel to the right of x. Since the jeep can carry only one load at a time, it must therefore cross the point x at least $k + 1$ times from the left. But between any two crossings from the left, there must be a crossing from the right, so there must be at least k crossings from the right. Then the jeep must arrive at the point x at least $2k + 1$ times, which is what the lemma asserts.

We now combine this result with Banach's formula and get

$$1 = (\text{path length between } x_{k+1} \text{ and } x_k)$$

$$= \int_{x_{k+1}}^{x_k} n(x)\,dx \geq (2k + 1)(x_k - x_{k+1}).$$

So $x_k - x_{k+1} \leq 1/(2^{k+1})$. Summing 0 to $f - 1$ implies

(3) $$\sum_{k=0}^{f-1} (x_k - x_{k+1}) = x_0 - x_f = d \leq 1 + \frac{1}{3} + \frac{1}{5} + \cdots + \frac{1}{2f - 1},$$

which gives an upper bound for $d(f)$. It remains to be shown only that this bound can be achieved, and this is easily done by induction. The formula is clearly correct when $f = 1$. Suppose now we are allowed $f + 1$ loads. Then let the jeep take $f + 1$ loads to the point $1/(2f + 1)$. This will involve $f + 1$ outward trips and f return trips, hence $2f + 1$ trips of length $1/(2f + 1)$. Therefore a total of one load is consumed, leaving f loads deposited at the point $1/(2f + 1)$. By the induction hypothesis, formula (1) holds from this point on, completing the proof.

For the case where f is not an integer, the same type of argument shows that

(4) $$d(f) = 1 + \frac{1}{3} + \cdots + \frac{1}{2[f] - 1} + \frac{\{f\}}{2[f] - 1},$$

where $[f]$ and $\{f\}$ are the integral and fractional parts of f, respectively. In other words, one simply interpolates linearly between integral values of f. The graph of $d(f)$ is plotted in Figure A2.3.

Since the odd harmonic series diverges, it follows that a desert of any size can be crossed.

After finding the solution just given, I became aware of the similar one given in [4]. The use here of the Banach formula seems to tighten the argument somewhat and make the reasoning more transparent. Also, we do not need to assume a priori that there will be only a finite number of depots. It is at least conceivable that the optimal solution involves infinitely many deposits or even a sort of continuous smear of fuel spread out along the route. One could no doubt formulate a very general problem in terms of measure theory.

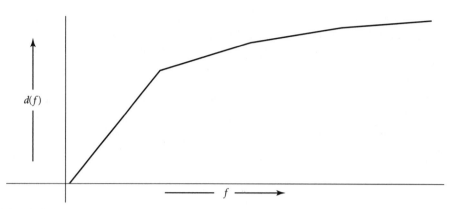

FIGURE A2.3.

The argument above shows, however—thanks again to Banach's formula—that this more general behavior could not give any improvement in fuel consumption.

Before leaving the single jeep, we consider the case in which the jeep is required to cross the desert and then return. For this case the arguments are the same as before, except that (2) becomes

(5) $$n(x) \geqq 2k + 2.$$

letting $d(f)$ be the longest possible round trip (e.g., twice the distance to the farthest point), we get the even simpler formula

(6) $$\tilde{d}(f) = 1 + \frac{1}{2} + \frac{1}{3} + \cdots + \frac{1}{f},$$

where as before equality is achieved. We note the familiar fact that a round trip is substantially cheaper than two one-way trips. In fact, comparing (6) with (1), we have

$$d(f) - \frac{1}{2}\tilde{d}(f) = 1 - \frac{1}{2} + \frac{1}{3} - \frac{1}{4} + \cdots - \frac{1}{2f},$$

but this is bounded by $\sum_{n=1}^{\infty} (-1)^{n+1}(1/n) = \log 2$. Thus, for a long distance, the increase in fuel cost for a round trip as against a one-way trip is negligible.

4. SEVERAL JEEPS

We turn now to the problem of m jeeps, and we shall compute the function $d_m(f)$ giving the most distant point that all jeeps can reach if they have f loads of fuel to share among them. In other words, the sum of the path lengths of all of the jeeps must not exceed f.

We proceed by defining the points x_k exactly as in the one-jeep case, as the point to the right of which the total combined path length of all jeeps is k. The analogue of Lemma 1, however, is slightly more complicated.

Lemma 2. *For any x in $[0,d]$, $n(x) \geqq m$. If $x < x_{m+r}$ $(r \geqq 0)$, then*

(7) $$n(x) \geqq m + 2r + 2.$$

Proof. The first assertion simply corresponds to the fact that all m jeeps are required to reach the point d. Concerning inequality (7), we see that more than $m + r$ loads must be transported beyond the point x, so this point must be crossed at least $m + r + 1$ times from the left. But since there are only m jeeps, to achieve $m + r + 1$ crossings from the left, there must be at least $r + 1$ crossings from the right, giving $m + 2r + 2$ crossings in all.

We proceed to calculate $d_m(f)$. Notice that for $f \leq m$, the problem is trivial and $d_m(f) = m/f$ (why?), so we assume that $f = m + s$, where $s > 0$. Again restricting ourselves to integral values of s, we claim that

$$d_m(f) = 1 + \frac{1}{m + 2} + \frac{1}{m + 4} + \cdots + \frac{1}{m + 2s}$$

(8)

$$= 1 + \frac{1}{m + 2} + \frac{1}{m + 4} + \cdots + \frac{1}{2f - m}.$$

Because the path length from x_{k+1} to x_k is 1,

$$1 = \int_{x_{k+1}}^{x_k} n(x)\,dx \geq m\,(x_k - x_{k+1}), \quad \text{for } k < m,$$

and

$$1 = \int_{x_{m+r+1}}^{x_{m+r}} n(x)\,dx \geq (m + 2r + 2)(x_{m+r} - x_{m+r+1}).$$

Therefore $x_k - x_{k+1} \leq 1/m$ for $k < m$, and

$$x_{m+r} - x_{m+r+1} \leq \frac{1}{m + 2r + 2}, \quad \text{for } r \geq 0.$$

Summing for $0 \leq k \leq m - 1$ and $0 \leq r \leq s - 1$ gives (8) as an inequality. Again an inductive proof shows that equality can be achieved. Assuming that the formula for $m + s$ loads of fuel is correct, we see that $m + s + 1$ loads can be moved to the point $1/(m + 2s + 2)$ by having one jeep make $s + 1$ round trips and then having all m of them make the one-way trip to this point. This will use up one load, so that $m + s$ loads are deposited, and the induction is continued. There are various ways in which this optimal journey can be performed. One way is to have one of the jeeps do all the work of setting up the various depots, while the others simply move outward without turning around, refueling as they go. It is, however, not possible to prescribe the same path for all jeeps, for if all jeeps followed the same route, then the function $n(x)$ would have to be divisible by m at all points, which clearly is not so in an optimal trip. We shall return to this point later.

To prove the result about increasing returns, we wish to compare the distance that one jeep travels on f loads with that which m jeeps travel on mf loads. So, we rewrite (8) for the case $s = m(f - 1)$ as follows:

$$d_m(mf) = 1 + \left(\frac{1}{m + 2} + \cdots + \frac{1}{3m} \right) + \left(\frac{1}{3m + 2} + \cdots + \frac{1}{5m} \right) + \cdots$$

(9)

$$+ \left(\frac{1}{2(f - 3)m + 2} + \cdots + \frac{1}{(2f - 1)m} \right),$$

where each term in parentheses contains m summands and there are f terms in all. We may also rewrite (3) as

(10)
$$d(f) = 1 + \left(\frac{1}{3m} + \frac{1}{3m} + \cdots + \frac{1}{3m} \right) + \left(\frac{1}{5m} + \cdots + \frac{1}{5m} \right) + \cdots$$
$$+ \left(\frac{1}{(2f-1)m} + \cdots + \frac{1}{(2f-1)m} \right),$$

where there are also m summands in each term in parentheses. A term by term comparison between (9) and (10) shows the advantage of using many jeeps. For the special case $m = 2$, we get

$$d_2(2f) - d(f) = \left(\frac{1}{4} - \frac{1}{6} \right) + \left(\frac{1}{8} - \frac{1}{10} \right) + \cdots + \left(\frac{1}{(4f-4)} - \frac{1}{(4f-2)} \right)$$
$$= \frac{1}{2} \left(\frac{1}{2} - \frac{1}{3} + \frac{1}{4} - \frac{1}{5} + \cdots + \frac{1}{(2f-2)} - \frac{1}{(2f-1)} \right).$$

Thus the longer the trip, the greater the saving. On the other hand, since the terms in parentheses above are part of a convergent series, it follows that the total amount saved remains bounded as the trip gets longer.

For the round-trip problem, the formula is again simpler. Let $\tilde{d}_m(f)$ be the round-trip distance that m jeeps can travel on f loads. Then

(11)
$$\tilde{d}_m(mf) = 1 + \frac{1}{m+1} + \frac{1}{m+2} + \cdots + \frac{1}{mf}.$$

So again, the more jeeps there are, the further out they can travel on the same amount of fuel, or, returning to the original problem the less fuel per jeep is needed to reach a preassigned point. But there is a limit to the advantage one obtains in increasing the number of jeeps, for as m approaches infinity, a standard calculation (comparison with $\int dx/x$) shows that

(12)
$$\lim_{m \to \infty} \tilde{d}_m(mf) = 1 + \log f.$$

So no matter how may jeeps there are, one cannot go further than $1 + \log f$ on f loads of fuel per jeep. In terms of the original problem, we get in the limit

(13)
$$f = e^{d-1},$$

so the amount of fuel needed increases exponentially with the length of the desert, as one might have guessed.

The multijeep round-trip problem can be interpreted in another way. Instead of thinking of m jeeps, each making a round trip, one may consider a single jeep that must make round trips, say on m successive days. The solution here is the same as for the m jeep problem, with the consequent saving of fuel. In fact, on the first day, the jeep can set up all the depots it will need for the following days, as one easily sees.

5. SOME FINAL REMARKS AND QUESTIONS

The last paragraph above raises an interesting question. Suppose, instead of being concerned about m round trips across the desert, one has decided to go into the desert-crossing business and plans to make a round-trip daily into the indefinite future. What sort of routine should one then use? As we have just seen, it would be uneconomical to use the single round-trip routine each day. Similarly, repeated use of the m-day routine is inferior to using the $(m + r)$-day routine, so that no periodic program of this sort could be optimal. On the other hand, if one is to consider programs that are not periodic, it is no longer clear what one means by an optimal program. In any case, it appears that the problem of finding the best "steady-state" routine has no exact solution at all, so that in practice one would have to settle for a routine that was "almost optimal."

There are many other jeep problems one can think of. Helmer [5] has considered some fairly complicated situations in which the number of depots one is allowed to establish is limited. To get a feeling for this sort of problem, the reader might look at the problem of crossing desert of length 2 when only 3 intermediate depots are permitted.

An apparently simple question is the round-trip problem in which fuel is available at both ends of the desert, but I must confess with embarrassment that I have not been able to find the solution. It is not hard to see that one can do at least as well in this case as in the case of two jeeps making one-way trips, but it may be possible to do better. The difficulty here, as with many optimization problems, is that there does not appear to be any simple way to determine whether or not a given solution is optimal. The upper bound given by Banach's formula does not seem to be available for this case. I put this problem forward as a challenge to jeepologists.

I conclude with some historical remarks. Shortly after the publication of Fine's solution, Phipps [3] derived the same result by arguing that the single-jeep problem is equivalent to a problem involving a convoy of jeeps that travel together, some being used to refuel others. The solution to the convoy problem is very simple, but the argument that this problem is equivalent to the original problem does not seem to be quite complete. Using this equivalence, however, one also easily derives the result given here on increasing returns. Finally, there is a feeling among many people that the jeep problem can be solved by the functional equation method of dynamic programming. In fact, the problem occurs as an exercise in Bellman's book [6], but the solution is not given there, and I know of no way of solving the problem by this method.

6. ADDENDUM

It has been pointed out to me that if one accepts the convoy equivalence of Phipps, then dynamic programming can be used (see, e.g., [7]). However, for the convoy problem, the solution is almost obvious anyway. Imagine that f jeeps set out. Since all but the last must return home, we suppose that $(f - 1)$ of them consume fuel at the rate 2 and the last one, the "red, white, and blue" jeep that will make the final crossing, consumes at the rate 1. Thus, initially the convoy is consuming at the rate $(2f - 1)$, which it does until one load has been consumed, after which the first jeep is abandoned. The remainder then go on consuming at the rate $2f - 3$, etc.

I should also remark that for Helmer's variation in which the number of depots is specified, dynamic programming does seem to be the appropriate tool.

REFERENCES

1. N. J. Fine, The jeep problem, American Mathematical Monthly 54, 24–31 (1947).
2. S. Banach, Sur les lignes rectifiables et les surfaces dont l'aire est finie, Fundamenta Mathematicae VII, 225–236 (1925).
3. C. G. Phipps, The jeep problem, A more general solution, American Mathematical Monthly 54, 458–462 (1947).
4. G. C. Alway, Crossing the desert, Math. Gazette 41, 209 (1957).
5. O. Helmer, A problem in logistics. The jeep problem, Project Rand Report No. Ra 15015, Dec. 1947.
6. R. Bellman, Dynamic Programming, Princeton University Press, 1955, p. 103 ex. 54–55.
7. J. N. Franklin, The range of a fleet of aircraft, J. Soc. Indust. Appl. Math. 8, 541–548 (1960).

Nineteen Problems in Elementary Geometry

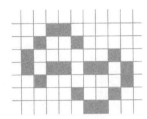

ARMANDO MACHADO

Several years ago somebody posed a problem in elementary geometry that seemed quite innocent but that resisted the more obvious approaches. It must be rather well known, as it appears repeatedly in mathematical circles. In Figure A3.1, we have an isosceles triangle with $\lambda = 20°$, $\alpha = 60°$, and $\beta = 50°$, and we must look for the values of γ and δ.

After the more obvious calculations, we obtain most of the angles missing in the figure but not the ones we want. All we conclude is that $\gamma + \delta = \alpha + \beta = 110°$. At this point I felt that although the problem was clearly well posed, there was perhaps no solution by the methods of elementary geometry. I then used some trigonometry and a pocket calculator to determine γ and δ. To my astonishment, the values were quite nice, namely, $\gamma = 80°$ and $\delta = 30°$. With such values there should exist an elementary solution! I remember that I used the same λ and a second pair of values for α and β (I don't know which any more) and that I again obtained integer values for γ and δ. I was beginning to believe that this was a general phenomenon, but a third trial told me that it was not so. For

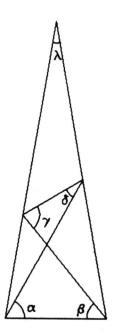

FIGURE A3.1.

TABLE A3.1.				
λ	α	β	γ	δ
20	50	20	60	10
20	50	40	60	30
20	60	30	80	10
20	60	50	80	30
20	70	50	110	10
20	70	60	110	20

example, with $\lambda = 20°$, $\alpha = 20°$, and $\beta = 70°$, we obtain $\gamma = 2°.12201$ and $\delta = 87°.87799$. ... I tried all the multiples of $10°$ for λ, α, and β, ignoring trivial cases, like those where $\alpha = \beta$, as well as those that were symmetric to previous ones. The only data that gave nice values for γ and δ were those in Table A3.1. (From now on, I will omit the degree symbol.)

The natural conjecture was that in all these cases there exists an elementary solution to the problem, one that does not involve trigonometric formulas. Indeed, this was verified by one of my colleagues, Margarita Ramalho. The interesting phenomenon was that for each case one had to present a different proof, a situation not very usual in mathematics. For example, the first two cases are quite trivial, although different from one another, and the fourth, the original one, can be solved by superposing its figure with that of the first case, mirrored in the vertical axis, and remarking that one equilateral and several isosceles triangles show up.

Recently, I was playing around with the mathematical software *Mathematica* when somebody raised the same problem. I decided to use this software to look for other initial data that could lead to elementary solutions. I asked *Mathematica* to try every integer value of λ and every integer or half-integral value of α and β, choosing the cases where the corresponding values of γ and δ were integers or half-integers. This experiment led to many more candidates for an elementary solution. Following an idea of David Gale, other candidates showed up, involving 7 as a divisor of the right angle. This led eventually to Table A3.2.

TABLE A3.2

Problem	λ	α	β	γ	δ
1A	60/7	390/7	150/7	480/7	60/7
1B	60/7	390/7	330/7	480/7	240/7
2A	12	42	18	48	12
2B	12	42	30	48	24
3A	12	48	12	54	6
3B	12	48	42	54	36
4A	12	57	33	75	15
4B	12	57	42	75	24
5A	12	66	42	96	12
5B	12	66	54	96	24
6A	12	69	21	87	3
6B	12	69	66	87	48
7A	12	72	42	108	6
7B	12	72	66	108	30
8A	20	50	20	60	10
8B	20	50	40	60	30
9A	20	60	30	80	10
9B	20	60	50	80	30
10A	20	65	25	85	5
10B	20	65	60	85	40
11A	20	70	50	110	10
11B	20	70	60	110	20
12	36	54	36	72	18
13A	45	45	15	52.5	7.5
13B	45	45	37.5	52.5	30
14A	360/7	240/7	120/7	270/7	90/7
14B	360/7	240/7	150/7	270/7	120/7
15A	360/7	345/7	150/7	435/7	60/7
15B	360/7	345/7	285/7	435/7	195/7
16A	72	39	21	48	12
16B	72	39	27	48	18
17A	72	42	24	54	12
17B	72	42	30	54	18
18A	72	48	24	66	6
18B	72	48	42	66	24
19A	72	51	39	81	9
19B	72	51	42	81	12
20A	120	24	12	30	6
20B	120	24	18	30	12

To be strict, not all of these cases are different. Cases 3A, 8A, and 12 all have the same solution and are special cases of a series where $0 < \lambda < 60$ is arbitrary, $\alpha = 45 + \lambda/4$, $\beta = \lambda, \gamma = 45 + 3\lambda/4$, and $\delta = \lambda/2$. In the same way, cases 3B, 8B, and 12 admit the same solution and are special cases of a series where $0 < \lambda < 60$ is arbitrary, $\alpha = 45 + \lambda/4$, $\beta = 45 - \lambda/4, \gamma = 45 + 3\lambda/4$, and $\delta = 45 - 3\lambda/4$. Thus Table A3.2 lists 36 problems with possibly different solutions. The number of different problems can be further reduced. From Table A3.2 a kind of duality is very easily discovered. To each problem $(\lambda, \alpha, \beta, \gamma, \delta)$ we can associate a dual $(\lambda', \alpha', \beta', \gamma', \delta')$, with $\lambda' = \lambda, \alpha' = \alpha$, $\beta' = \alpha - \delta, \gamma' = \gamma$, and $\delta' = \alpha - \beta$ (problem 12 is self-dual). We thus have 18 problems once we prove this duality, this proof being the nineteenth problem referred to in the title. It can be solved by superposing the figures corresponding to $(\lambda, \alpha, \beta, \gamma, \delta)$ and $(\lambda', \alpha', \beta', \gamma', \delta')$, after mirroring one of them in the vertical axis and then applying a special case of Pappus's theorem from projective geometry. I did not solve each one of these new cases by elementary methods (in fact, the answer given by the computer is not a true solution, as will be evident below). This is perhaps a good activity for dead periods in academic life, such as boring meetings of the academic staff.

This is also a useful illustration of the care that one must take with computer applications in mathematics. I was using the software *Mathematica* with its default precision, which gives about six correct digits, and my program tested whether or not a number was an integer or half-integer by a method that amounts to looking at the first five decimal digits of its double. Beside the cases in Table A3.2, the computer also proposed those in Table A3.3.

There are several strange things in Table A3.3. First, there are three values for δ that are almost integers or half-integers but are not exactly so. Second, six dual problems are missing. This could be caused by some roundoff errors, so I used the ability of *Mathematica* to work with an arbitrary number of digits and tested each of the values in Tables A3.2 and A3.3 with 50-digit approximation. Every value in Table A3.2 remained correct, but all

TABLE A3.3				
λ	α	β	γ	δ
5	74.5	49	117	6.5
5	74.5	68	117	25.5
8	33.5	5.5	35	4
8	33.5	29.5	35	28
23	46	15.5	53	8.5
23	46	37.5	53	30.5
35	33.5	18	37	14.5
39	33.5	11.5	36.5	8.5
39	33.5	25	36.5	22
41	67	13	79.5	0.500003
59	33.5	4	35	2.5
59	33.5	31	35	29.5
61	32.5	7.5	35	5
61	32.5	27.5	35	25
67	32	4	33.5	2.5
67	32	29.5	33.5	28
68	46	28.5	62	12.5
68	46	33.5	62	17.5
68	55.5	41.5	95.5	1.5
77	49.5	14	62.5	0.999998
78	44	25.5	60.5	9
97	23	1.5	23.5	0.999998
97	23	22	23.5	21.5
129	21.5	7	26	2.5

the values in Table A3.3, as well as their missing duals, appeared only as approximate integers or half-integers, although with a very good approximation. For example, the first value for δ in Table A3.3 becomes

6.50000 16063 95883 45352 05203 60429 84883 68773 . . . .

At that moment I was astonished by what seemed an incredible coincidence: Several results that were half-integers within five decimals but were not real half-integers. In fact there was no occasion for astonishment: The computer had tried about 950,000 triples, many more than the number of groups of five digits.

The Truth and Nothing But the Truth

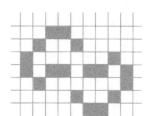

As far as the laws of mathematics refer to reality, they are not certain; and as far as they are certain, they do not refer to reality.

Albert Einstein

In this chapter I write about a particular aspect of mathematics. The exposition, however, is deliberately informal and nonrigorous. Readers who are so inclined will find plenty of nits to pick, but if I were to proceed throughout with complete rigor, it would impede the flow of ideas. So I hope the reader will not begrudge me some limited use of poetic license.

When I started writing this chapter, my idea was to give a rough exposition, mainly via examples, of a controversial area in the philosophy of mathematics without taking sides myself. But in the course of writing, I found myself reaching positions on at least some of the issues. I decided to present my conclusions in the final paragraphs. I realize that others have thought about these matters over the years, probably in much greater depth than I, and if some of them disagree with my conclusions, I know that I and perhaps other readers would enjoy reading their rejoinders.

THE MEANING OF MATHEMATICAL RESULTS

Mathematics is distinguished among academic disciplines by the property that it alone establishes facts with absolute certainty. Once a theorem has been correctly proved, no "on the other hand" is possible. Thus, for example, every positive integer can be written as the sum of at most four squares, period. This noncontroversial nature of the results of mathematics extends, it seems to me, to other aspects of mathematical culture. Mathematicians generally agree, for example, as to which results in the subject are important, e.g, the fundamental theorem of algebra, Cauchy's theorem, the central limit theorem, Gödel's incompleteness theorem (you can probably make up your own list of the top ten theorems of all time). There is also a fair consensus within the mathematical community as to which contemporary mathematicians are doing the most significant and exciting work. It is therefore almost refreshing to find in this sea of harmonious concord one facet of the subject on which mathematicians sharply disagree. The dispute is not about the validity or profoundity but rather about the *meaning* of mathematical results and, more specifically, how they relate to reality.

Let me begin with some illustrations. We know that it is impossible to tour the city of Königsberg in such a way as to cross each of its seven bridges exactly once, and the way we arrive at this knowledge is by reasoning that is surely mathematical. We also know that it is impossible to trisect an arbitrary angle using only ruler and compass, and the reasoning here is considerably more profound and subtle. We know that in a right triangle $a^2 + b^2 = c^2$ (this is on my top-ten list), but until recently we did not know whether the analogue of this equation has integral solutions when the exponent is greater than two. Most of us believed, however, that this was a matter of our ignorance and *in reality* such integral solutions did or did not exist.

On the other hand, some mathematical results clearly have no counterpart in reality. It is illuminating to contrast a pair of well-known decomposition theorems. The first is the Bolyai–Gerwin theorem, which says that if two polygons P and P' have the same area, than we can cut up P into a finite number of pieces (triangles in fact) and reassemble them to get P'. This theorem certainly has a real counterpart, say in the construction of jigsaw puzzles. I have a computer program that allows the user to place a more or less arbitrary triangle on the face of the monitor. The machine then chops it up and reassembles the pieces to form a square of the same area. You can't get much more real than that. By contrast, the Banach–Tarski theorem gives the same conclusion as Bolyai–Gerwin for any two balls B and B' of *different volumes* in 3-space. I think most would agree that this theorem has no counterpart in the real world (much less an implementing computer program). What I hope to do in this essay is to explore the "reality gap" between results like Bolyai–Gerwin on the one hand and Banach–Tarski on the other, because, as I have stated, there is wide difference of opinion on these matters. I will begin by quoting two such opinions. The subject, as one might have guessed, is set theory.

Raymond Smullyan is lecturing on the continuum hypothesis to a group of nonexperts.[1] Having explained that both the hypothesis and its negation are now known to be consistent with the standard axioms of set theory, he says, "Now, some very formalistic mathematicians say this means that the continuum hypothesis is neither true nor false. Well, I don't buy that and neither would most—many—mathematicians. We want to know whether the continuum hypothesis is *true* or *not*" [my italics]. His position is that our present tools, say Zermelo–Frankel set theory, are insufficient to let us know whether

---

[1]From the Public Broadcasting System NOVA program, "*Mathematical Mystery Tour.*"

or not there are cardinals between aleph null and the cardinality of continuum, but he is in no doubt that either there are or there aren't.

Saunders Mac Lane [2] has quite a different point of view. He says, "the belief of logicians in Zermelo–Frankel set theory is apparently based on a sort of Platonic imagination that there is out there in the real world some universe of sets to which their axioms will apply. ... I submit that the Platonic notion of a real universe of sets is absolute balderdash."[2]

Without taking sides on the issue of the continuum hypothesis (yet), I believe Smullyan is wrong on one point. It is not true that most, or even many, mathematicians believe that the continuum hypothesis is either true or false. Indeed, I have yet to find a mathematician, aside from those working in set theory, who espouses this view. As one person put it, we know that we can have both Euclidean and non-Euclidean geometry. Now we see that there are different kinds of set theory—and what's wrong with that? On the other hand, some, though not all, of the set theorists to whom I have talked take the Platonic view. In my limited sample two of them believed that the hypothesis was probably false, as did Gödel. In fact, toward the end of his life, Gödel had come to the view that the cardinality of the continuum was probably aleph two, and there are even some writings of his that attempted to formulate a system that would have this as a consequence. In recent times, I am told, some of the younger set theorists believe that the CH is true. There are a fair number of undecideds, and also, even among set theorists, many who consider the question meaningless.

TRUE, FALSE, OR MEANINGLESS

We are dealing of course with the famous law of the excluded middle. For clarity let us say that a statement is *determinate* if it satisfies the law of the excluded middle and *indeterminate* if it does not. An example of a statement I think most people would consider determinate is, "Cleopatra had blood type A." Although we will surely never know whether the statement is true or false, it is clearly one or the other. On the other hand, consider the statement, "If Gauss had never been born, someone else would have proved the fundamental theorem of algebra before the end of the nineteenth century." Although it is clear what is being asserted here, it is not appropriate to ask whether the statement is true or false. The statement is rather a manner of speaking. Shall we then consider all such conditional statements to be indeterminate? What about, "If I hadn't sent you that letter you would not have received it?"

My point is that even in ordinary language it is not always clear whether a particular statement is determinate or indeterminate. As a final illustration, consider, "The sun will continue to exist for at least 10 billion years after all life on Earth has disappeared." I'm not sure I know where I stand on that one myself. Physics is such a crazy subject that perhaps the very notion of time, that is, years, is indeterminate!

Getting back to mathematics and mathematicians, there seems to be a broad spectrum of beliefs on the excluded-middle issue. On the extreme right are those who reject using the law of the excluded middle in any argument, while at the left we have the Pla-

---

[2]Absolute balderdash is presumably a strong form of balderdash by analogy with absolute and ordinary convergence.

tonists, like the set theorists referred to earlier, who believe that the law holds in great generality. For us "middle"-of-the-roaders, things like Fermat's last theorem are either true or false.* Not so for the right-wingers, who feel it may well be indeterminate. On the other hand, the mainstream people believe that the continuum hypothesis is indeterminate. Why?

Let us examine this question in a specific context and start from a specific set, namely, the natural numbers N. Mainstream people believe that there is such a set, which has many interesting properties, one of which is that in N Fermat's last theorem either is or is not true. Next we go on to the set variously denoted by $p(N)$ or 2^N, which is the set consisting of all subsets of N. The continuum hypothesis asserts that every nondenumerable subset of $p(N)$ is equinumerous with $p(N)$. So the question is, Why does the "mathematician on the street" consider this statement to be indeterminate, while Fermat's theorem is determinate? By chance I happened to be present on the street and overheard the following dialogue.

> *Plato.* Why do you believe the continuum hypothesis is neither true nor false?
>
> *Mathematician on the Street.* Well, Gödel showed that it was consistent and Cohen showed it was independent of the rest of set theory.
>
> *Plato.* What do you mean, "the rest of set theory?"
>
> *MoS.* Well, there's something called Zermelo–Frankel set theory. It's a collection of axioms, I believe.
>
> *Plato.* Do you know what these axioms are?
>
> *MoS.* Well, no, not really.
>
> *Plato.* Have you worked through the proofs of Gödel and Cohen?
>
> *MoS.* No. That's not my area. I work in number theory. I'm trying to prove Fermat's last theorem.
>
> *Plato* Oh, number theory! So you work with the natural numbers.
>
> *MoS.* All the time.
>
> *Plato.* And do you sometimes consider sequences of natural numbers?
>
> *MoS.* Quite often.
>
> *Plato.* Do you ever think about *all* sequences of natural numbers?
>
> *MoS.* Once in a while. I usually give it the product topology and denote it by Σ.
>
> *Plato.* I see. And I take it then that this Σ is quite a well-defined object in your mind. And do you sometimes look at subsets of Σ?

---

*This was written six years before Fermats theorem was proved.

MoS. Oh yes.

Plato. And would it be fair to say that you feel quite comfortable working with such subsets?

MoS. Sure. Some of them are open, some are closed, some are measurable, some are not. It's very interesting.

Plato. And when you work with these various kinds of subsets, you are always working in the framework of the Zemelo–Frankel axioms?

MoS. Oh no, certainly not. As I've already said, I don't even know precisely what these axioms are, and in any case I don't think such formal axioms play any role in my work or that of most other working mathematicians.

Plato. Well, now you have me really confused. You tell me you feel comfortable working with general subsets of Σ, yet when I ask you a straightforward question about the cardinality of such subsets, you reply that the answer is indeterminate, neither true nor false. And when I ask why you feel that way, you tell me that you've heard via the grapevine that a couple of famous mathematicians have shown that the hypothesis can neither be proved nor disproved from a particular set of axioms—axioms that as far as you're concerned "don't play any role" in the work you are doing anyway. Isn't it possible that this is just a shortcoming of those axioms and not an indeterminacy of the facts?

MoS. I seem to be getting a headache.

Plato. I have no further questions.

Having heard all this and being a street person myself, I too felt some confusion. To exclude or not to exclude the middle, that was the question. I began to think about the right-wing nonexcluders. Surely even they must agree that some statements are determinate. For example, what about the statement

(1)   The 100^{100}th decimal digit of π is a 7.

Now, there is an algorithm, actually many, for settling this question (and in fact some of them are being actively executed at the present time; see [3]). Nevertheless, it seems certain that we will never know the answer to (1) because the time required to get it probably exceeds the expected duration of life on earth (if not, add on a few more zeros in the exponent). But in principle the calculation could be carried out, and I don't believe anyone has proposed a theory that maintains that whether or not a statement is determinate depends on the length of time required to verify it.

Next I thought of the following statement:

(2)   The decimal expansion of π contains a string 100^{100} consecutive 7s.

At this point the right-wingers might claim that the statement may be indeterminate, for while there is a finite (though lengthy) procedure that would in principle verify the statement if it is true, there is no such procedure in case the statement is false. This situation, with signs reversed, is the same as that of the Fermat problem. If Fermat's theorem is false, a finite procedure will confirm this, namely, find suitable numbers a, b, c, and n, and perform the required arithmetic. If the theorem is true, no such finite procedure exists.

Then I went one step further and considered the following:

(3)   The decimal expansion of π contains an infinite number of 7s.

At first glance this would seem like a more tractable statement than (2) (there is very strong "empirical evidence" that the statement is true; again, see [3]). But from another point of view, statement (3) is a better candidate for indeterminacy than (2). In this case neither the truth nor the falsehood of the proposition can be verified by finite calculation, as a moment's reflection will show. (The analogous situation among the classical unsolved problems is the twin prime conjecture.)

Although we will almost certainly never know whether (1) is true, we may conceivably someday know the answers to (2) and (3). Indeed, there is the well-known conjecture that π is a *normal number,* meaning that all its digits to base 10 or any other base b are uniformly distributed, and more generally, every consecutive sequence of digits of length n occurs with asymptotic density b^{-n}. If someone should prove this, then statements (2) and (3) would be obvious consequences. So let the next statement be

(4) π is a normal number.

Now, I would like to speculate that (4) will *never* be either proved or disproved. In fact, let me suggest that perhaps there *is* no proof or disproof of (4) in exactly the same sense that there is no proof or disproof of the continuum hypothesis; namely (excuse the repetition), it is not provable from standard Zermelo–Frankel set theory. This may strike some people as far-fetched. It's one thing, they might say, to show that a nebulous proposition about the continuum is unprovable, but when it comes to concrete questions about the density of subsequences in a well-defined sequence, one would expect that such questions ought to be resolvable by standard though possibly very involved arguments. In response to this I would point out that some statements much more "down to earth" than (4) are *known* to be unprovable from Zermelo–Frankel. Perhaps the best example is a by-product of the work on Hilbert's tenth problem. A specific polynomial P with integer coefficients and, I believe, nine variables has the property that there is no proof or disproof (from Zermelo–Frankel) that this polynomial has a root in the natural numbers, i.e., a 9-tuple of positive integers $n = (n_1, \ldots, n_9)$ such that $P(n) = 0$. This problem is, in one sense at least, simpler than the Fermat problem. The Fermat equation $a^n + b^n - c^n = 0$ is more complicated than a polynomial because variables occur as exponents.

So now my question is this: Suppose someone should actually prove that (4) could not be proved or disproved from Zermelo–Frankel.[3] Would you then conclude that it was neither true nor false? If your answer is yes, then let me ask the same question about (3).

---

[3]This possibility is admittedly highly hypothetical. The known techniques for proving undecidability do not seem appropriate for statements like (3) and (4), but one can imagine that some as yet unknown methods would give the result.

Surely, I thought, my friend MoS believes that the expansion of π contains either a finite or an infinite number of 7s. Would he change his mind if someone showed that the statement was unprovable? I decided to ask him this when I ran into him on the street a few days later. I could see he was feeling well again, having recovered from his earlier encounter with Plato. I put the above question to him and he responded right away.

"Of course, there are either a finite or infinite number of 7s in the expansion of π regardless of whether or not someone proves it to be unprovable. This is quite a different story from the continuum hypothesis. As what's-his-name said, God gave us the natural numbers. All the rest, like sets and cardinal numbers, are the creation of people like, in this case, Cantor and Zermelo."

(When I mentioned that π was not a natural number he pointed out rightly that this was beside the point. What we were really talking about were algorithms that for each *natural number n* produced a digit. The fact that this particular algorithm came from expanding π was clearly irrelevant.) I persisted and asked whether the fact that (3) could not be proved from the Zermelo–Frankel axioms would raise any questions in his mind. He replied, with some impatience, "Axioms, schmaxioms! We know there are lots of things that can't be proved from those bloody axioms. Plato was right about that."

I wasn't going to let him off the hook yet, and so I asked "And what if it were shown that the normality of π is undecidable in Zermelo–Frankel? That involves statements about asymptotic densities of subsequences and so on. Did God give us that, too, or was it human invention?"

He looked at me for a moment and then said, "I'll have to get back to you on that." Unfortunately, I haven't run into him since.

Anyway, here was more food for thought. Where should one draw the line? At what point does God leave off and man take over? Where is the boundary between reality and balderdash?

Reality

I started experiencing some head symptoms myself at this point and decided to let the matter rest for a while, but some days later I happened to be reading Donald Knuth's article "What Is a Random Sequence?" [1], and the question came up all over again. I thought back to statement (4) again. It appears that π satisfies conditions much stronger than merely being normal. For example, if one looks at the digits $d_1, d_4, \ldots, d_{n^2}, \ldots$, they too seem to be uniformly distributed. No doubt the same is true for the subsequence (d_{n^3}). We have a lot of data by now (at last count we knew more than 2 billion digits). This suggests that we should strengthen the definition and talk about *supernormal* numbers as those that are not only normal but such that are "describable" substances are normal. (Obviously we cannot require that *all* subsequences be normal because then, given any number, one could choose the subsequence as those terms where the digit was 7, and there would be no supernormal numbers.)

Fortunately, the word *describable,* as probably most mathematicians and all computer scientists know by now, can be given a precise definition. The correct technical term is *recursive.* One should think of such sequences as those for which there is an algorithm for calculating each term. The crucial fact is that there are only countably many such sequences because there are only countably many algorithms. Let us then call a number *supernormal* if all of its describable subsequences are normal. One thing is immediately clear. The number π cannot possibly be supernormal because obviously the places where the expansion of π has the digit 7 is a describable subsequence (I just described it) and the corresponding number, $0.777\ldots$, is highly abnormal. Intuitively, we think of a supernormal number in base 2 as one obtained by repeated tosses of a fair coin. It would surely be surprising, would it not, if in 16 million tosses the coin fell heads whenever the corresponding decimal digit of π was a 7. The point is that supernormal numbers, *if they exist*, must be "indescribable." But they *do* exist. In fact, the set of *non*supernormal numbers is a set of measure zero! Further, the set of supernormal numbers is a nice set, an $F_{\sigma\delta}$ to be exact. The proof of this result, which is essentially due to E. Borel, is not difficult and involves nothing more profound than a form of the law of large numbers.

I find this a very strange result, not because of the mathematics, but because of what it seems to mean. If we think of flipping the true coin repeatedly, the sequence we get will almost surely correspond to a supernormal number and hence, by definition, a sequence that is impossible to describe. The mathematics is clear, but the question is, in the words of Mac Lane, Do such sequences "exist in the real world" or are they just figments of our mathematical imagination? I don't know what Plato would say at this point, but I have reached some conclusions of my own.

I believe the continuum hypothesis is neither true nor false, but not because of the results of Gödel and Cohen, though these results strongly reinforce my conviction, and I would not have arrived at it without them. To explain my position, let us go back to the specific context of the natural numbers N and the set $p(N)$ of all subsets of N—repeat, *all* subsets of N. That "all" is the joker. In "reality," none of us has ever seen an infinite set. We've seen plenty of finite sets, and we certainly know what we mean by all subsets of such a set. Most mathematicians also *think* they know what they mean by the set of all subsets of *any* set, finite or infinite. Even my friend MoS is among them with his set Σ. But wait! When we see the word *all* in connection with set theory, a red light should go on in our minds. Cantor believed it was acceptable to think of the set of *all* sets, and we know the kind of difficulty this led to. Up to the present the set of all subsets of a set has not led to any contradictions, but my thesis is that it marks the point at which mathematics loses contact with reality.

The left-wing set theorists, and this included the great Gödel, believe there exists *in reality* a specific set that is the set of all subsets of the natural numbers (and some of them believe a lot more). Well, then, either they know something I don't know or they're kidding themselves, extrapolating from something that is clear and verifiable in the finite case to the infinite case where, in my opinion, the concept is probably meaningless, just as meaningless as asking what would have happened if Gauss had never been born. This is not to say that the concept of $p(S)$ is not important in many branches of mathematics. I'm sure I've used it many times. All I'm saying is that at that point we're talking mathematics, not reality. So for me the reason the continuum hypothesis is indeterminate is because the sets involved are indeterminate.

As to supernormal numbers, again we must start with the set of *all* real numbers (or *all* sequences of digits), throw away a bunch of them, and consider those left. Because the set of all real numbers makes me nervous, I certainly don't place much stock in the existence of this weird subset. Looking at it the other way, we are talking about the heads–tails sequences we would get almost surely if someone flipped a fair coin an infinite number of times. In other words, we are talking about what would happen if something that can't happen happened. I find this even *less* acceptable than talking about what would have happened if Gauss had not been born.

Unfortunately, this is not the end of the story. I still have to face the problem of the infinite number of 7s in the expansion of π. It comes down to this: What are we doing when we write down expressions like, "three point one four one five nine dot dot dot"? I feel quite comfortable with three, one, four, five, nine, and point. It's the "dot dot dot" that bothers me. I think I know what I mean by it—but do I? (and where did I put that aspirin bottle?)

So where does all this place me in the spectrum of beliefs about the law of the excluded middle? Somewhat right of center I would say. Perhaps I should call myself a fuzzy-minded conservative. Anyway, that's the way I feel right now. Of course, all this could change by next week, especially if I should happen to run into Plato in the interim.

REFERENCES

1. D. E. Knuth, *The Art of Computer Programming*, Reading, Mass., Addison-Wesley, 1969, Vol. 2.
2. S. Mac Lane, "Are we all just specialists?" *Mathematical Intelligencer* 8, 4, 74–75 (1986).
3. S. Wagon, "Is π normal?" *Mathematical Intelligencer* 7, 3, 65–67 (1985).